BUILD A PERSONAL EARTH STATION FOR WORLDWIDE SATELLITE TV RECEPTION

Design, build, install, operate, and maintain your own TV reception unit!

BY ROBERT J. TRAISTER

TAB **TAB BOOKS Inc.**
BLUE RIDGE SUMMIT, PA. 17214

FIRST EDITION

SEVENTH PRINTING

Printed in the United States of America

Library of Congress Cataloging in Publication Data

Traister, Robert J.
    Build a personal earth station for worldwide
satellite TV reception.

    Includes index.
    1. Earth stations (Satellite telecommunication)
I. Title.
TK5104.T7   621.388′8   81-18279
ISBN 0-8306-0090-6   AACR2
ISBN 0-8306-1409-5 (pbk.)

Contents

Introduction   vii

1 The Television Age   1

2 Satellite Broadcasting   8
Transmitting Stations—Receiving Stations—Radio Waves—Line
of Sight Broadcasts—Skywave Communications—Space
Communications—Summary

3 The Personal Earth Station   25
The Satellite Receiver—A Summary of the Basic TVRO Earth
Station—The Enjoyment of a TVRO Earth Station—Questions
and Answers—Summary

4 Obtaining Surplus Components   54

5 Principles of TVRO Antennas   60
Waveguides—Horn Antennas—Paraboloid Reflectors—
Spherical Antennas—Parabolic/Spherical Comparison—Other
Types of Antennas—Summary

6 Commercially Available Antennas   85
Wilson Microwave Systems—Microwave Associates Communications TVRO Antennas—Skyview I Antenna—Winegard Earth Station Antennas—Satvision Model SV-11—Comtech 3-Meter
Antenna—Comtech 5-Meter Antenna—Summary

7 Commercially Available LNAS, Receivers, and Modulators   124
Avantek AWC- 4200 Series—Microwave Associates Communica-
tions Company LNA—Avantek LNA/Downconverter—Gillaspie
and Associates Model 7500—Vitalink Communications Corpora-
tion Model V-100—Downlink D-2X Receiver—Microdyne Corpo-
ration 1000 TVRN—HR-100 Satellite Receiver—Avantek AR1000
Receiver—Comtech Model 650 Receiver—ICM TV-4000 Satellite
Receiver—ICM TV-4400 Satellite Receiver—VR-3X Satellite
Receiver—VR-4XS Satellite Receiver—Brief Notes On Other
Receivers—Accessories—Summary

8 **TVRO Systems**   175

9 **Multiple Access Ground Stations**   184

10 **Satellite Location Techniques**   193

11 **A Commercial Cable Television System**   210

12 **Building Your Own**   219
    A TVRO Receiver Kit—Mixer Module—First I-f Amplifier—Second
    I-f Amplifier—Video Demodulator—Audio Demodulator—Power
    Supply—Control Console—Installation—Soldering—Receiver
    Alignment—Summary

13 **Specific Site Selection**   244

14 **The Earth Station Assembly**   254
    Site Determination—Equipment Selection—The Equipment
    Arrives—The Checkout—Problems—Calling the Manufacturer—
    Interference—Summary

Appendix A: **Satellite Video Systems Complete Listing of**   263
             All Programming Sources Available from the
             Satellites

Appendix B: **List of Geostationary Space Stations**   267
             by Orbital Positions

Appendix C: **Table of Artificial Satellites Launched in 1980** 270

Appendix D: **LNA Noise Temperature to Noise**   282
             Figure Conversion Chart

Appendix E: **Wind Pressure Table**   283

Appendix F: **Common Abbreviations**   284

Appendix G: **Noise Temperature to Noise Figure Table**   286

Appendix H: **Manufacturers and Suppliers of TVRO**   287
             Earth Station Components

Appendix I: **Satcom I Transponder Frequencies and**   290
             Polarization

Appendix J: Example of Manufacturer-Supplied   291
Antenna Pointing Angles for Specific
Satellites

Appendix K: The Frequency Spectrum   292

Appendix L: Preliminary EIRP Contour Map   293

Appendix M: Preliminary Carrier/Noise Versus EIRP   294
Chart

Index   295

This book is dedicated to the memory of the late William Cyrus Gilmore, Jr., my best friend. His contributions to the entertainment field as a radio/television manager and actor, along with his production of the U.S. Air Force "Serenade in Blue" program, touched many. His sense of humor and unparalleled story-telling abilities reached millions through the Ed Sullivan Show and as the Brooklyn cabbie in several television commercials. He will be remembered by many as the funniest comedian who ever lived and by those who knew him well as their finest friend.

Introduction

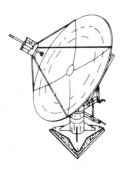

Television from space! That's what everyone seems to be talking about. Yes, now it's possible for nearly anyone to receive television broadcasts from a number of satellites in geostationary orbits nearly 23,000 miles above the Earth's equator. Can I use my present television receiver? Do I need a special license? How much will it cost? These and many other questions are being asked today by individuals who are interested in exploring this field. These are the same questions that are also asked by those people who would simply like to be able to receive the uninterrupted movies, 24-hour news, and religious and educational programs being offered only by satellite broadcasts.

Whether you are interested in experimenting with your own personal Earth station for satellite reception or in simply receiving additional channels on your home television receiver without having to resort to technical schooling or electronic tinkering, the TVRO (television receive-only) Earth station is probably right for you. No, this does not need to be a highly complex undertaking. Anyone with a small amount of do-it-yourself ability can assemble a TVRO Earth station. These stations are usually assembled from a kit of components supplied by the manufacturer. No soldering or complicated wiring is needed. In some ways, the setup is very similar to installing a component stereo system. The amplifier goes here, the speakers go there, the turntable and tape deck are positioned here, and all are connected by cables.

This is what building your own satellite TV Earth station is all about. It doesn't take long (usually less than a day), and the costs are not as excessive as they once were. Today, many persons can set up a completely operational Earth station to receive the programs from satellites for less than $2,500. This compares favorably with the price of many video cassette recorder/color camera outfits and may even be less than the price of many giant screen color television sets. And the price will continue to drop. More and more persons are installing their own TVRO Earth stations, and television programmers along with sponsors are responding to this interest by making available more and more programs and services to satellite broadcast viewers.

This book takes you through the basics of television reception as we have experienced it for many years, then moves on to satellite television broadcasts. It explains the basics of how the signal gets to your site, and then goes on to tell you what is involved in your own personal Earth station . . . how it's set up . . . how to find the satellites . . . security precautions . . . and all the information that's needed to understand this fascinating and imminently useful service.

Many Earth station component manufacturers were contacted in researching this book, and the book is filled with the latest information available on this ever-growing industry.

Whether you are an avid experimenter or a person who just likes to watch good television programs, a TVRO Earth station might be your next major appliance buy. It is hoped that this information will tell you all about what is available from these orbiting broadcast stations and that you will be as excited by what can be had in the comfort of your own living room as I am.

Chapter 1
The Television Age

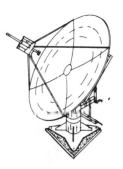

Most of us on Planet Earth have several things in common. One of these is the television age. We are all a part of it and here in the United States, we depend upon our personal television receivers for more than we consciously realize. Think of a world without television. Admittedly, it's pretty hard to conceive of the idea. This is how dependent we have become on this (now) not so modern electronic invention.

Certainly, there are those who will say that the American people would be a lot better off without the TV receiver. This is difficult to imagine, because through this electronic marvel, the entire world is at our fingertips. The television receiver brings information into every home. Information breeds knowledge if properly mated with the human mind. Sure, television viewing can be abused, but this fault lied with the viewer not with the device itself.

I can still remember the first television set I ever viewed. This occurred through a department store window, against which a youthful face was flatly pressed in order to get as close as possible to that miniature picture tube. And what wonders that tube displayed! Cartoon figures with huge heads and little bodies scurrying here and there in comic disarray! How many readers can still remember the cartoon drama (lasting all of three minutes) entitled "Barnyard Bunk", and starring the little farmer? The Mickey Mouse Club? Or even Winky Dink (starring Jack Barry)? Sure, these were all "kid shows", but in the middle fifties, I was just that, a pre-school kid.

Now, the education radicals will ask what good was derived from these senseless programs? I say, "bunk" (or should I say Barnyard Bunk?). These shows stimulated youthful imaginations which lead to other endeavors (some avid viewers even became writers).

Reflecting again on the past, the first television receiver my family purchased was about a year after my first contact with the device through the department store window. What a pleasure it was to return home and find white-coveralled men climbing to the top of our peaked roof, precariously erecting a mast with stacked, two-element, yagi antennas at the top. The immediate rush of tiny legs into the living room brought a grand sight to ever-widening eyes which threatened to burst from their sockets and fly straight for that large picture tube (21 inches). There on a hastily uncluttered table sat the biggest "TV" the world had ever known. Made by Dumont, if my memory serves me correctly, it boasted full reception of channels two through thirteen, had a massive black case, and must have weighed two hundred pounds!

The installation took a bit longer than anticipated. The workmen stayed until past 7:00 P.M. Labor was included in the price of the receiver, mast, and antenna, with the total cost running to nearly one hundred dollars. "What extravagance," my mother must have thought. That night, my father did the honors by cautiously flipping the control to the on position, amid the cheers of a gathering audience which consisted of half the neighborhood (children and adults alike). "Make it play! Make it play!" was an oft-heard cry.

And play it did. I vividly remember watching a documentary on what is now Saudi Arabia. Thinking back, the television announcer was wrong, because I distinctly remember pyramids along with the camels and the people, so the country was obviously Egypt. But this didn't matter then. Television was so new and wonderful that it didn't make any difference what was on, as long as it was something.

This last statement is not as ridiculous as it might sound. In those days, we didn't have cable television service. While our location was near Washington D.C., and Baltimore (close enough for television reception), TV stations didn't broadcast into the wee hours of the morning. Some of them didn't start to transmit until late morning or even early afternoon. Most were off the air by 10:00 in the evening. With these conditions, we were lucky to get an hour or so of viewing in each day.

But this was the peak of the television's growth from being an oddity to serving as a major appliance in the home. More channels started to become active as new stations went on the air. Others increased transmit power and were suddenly available to us. Often, the receiver was activated just to see what new station had sprung up from nowhere. Americans wanted television, and they sure got it!

The amateurish notions of many locally oriented programs began to take on a more professional air. We children saw the latest toys which were on sale, "Just in time for Christmas giving". We became aware that there were people on the face of the earth who did not look exactly like we did. We even discovered that there were all sorts of funny looking tubes which ran through our bodies and performed a myriad of services for us, without which life would be impossible.

But again, there were the cartoons. Those glorious animations which showed elves, dragons, dwarfs, fairy princesses, and ocassionally, real people. Buffalo Bob and Howdy Doody were daily treats. We cheered on the heroes and hissed the villains who never seemed to do any real harm. We learned right from wrong and that you were always punished in some form when you chose the wrong path. Let's face it, most of us received a large part of our ideas about life itself from television . . . in partnership with parental guidance. So, to say television is bad is also to say knowledge is bad. Knowledge is always good. What we choose to do with it is another thing altogether.

It is, I feel, totally appropriate to mention that the old Dumont which I devoured whenever possible as a child for the cartoon wonders it held was the same set that, ten years later, I watched the coverage of the John F. Kennedy assassination. I even saw his accused assassin, Lee Harvey Oswald, mortally wounded while being covered in a live broadcast. The same rapt attention was there as when viewing the cartoons years earlier, and I and millions like me were still learning. However, we were not overjoyed by what we saw. Rather, we were repelled by a true fact of life.

Television was becoming more and more true to life. Some of the entertainment or amusement aspects were beginning to fade away. Their place was taken by news, information, and definite realities. Ideal lifestyles were still presented, but this was evenly balanced by harsh events which were happening on a daily basis. Some will blame television for these disgusting realities, but I don't agree.

You are probably wondering at this point what these maudlin reflections and opinionated statements have to do with a book on satellite television Earth stations. Admittedly, it may be a rather unusual way to start a book; but it is vastly important to realize just why people are beginning to turn to increased coverage of the world and, indeed, outer space through the use of satellite transmissions. It all boils down to the importance we all place on information. Television is the most convenient method available to us today for the gathering and dissemination of information. It has irrevocably changed the world and certainly for the better. It continues to instigate changes.

Take our American political system, for example, and our methods of electing a President. Our individual votes do not directly elect a person to this high office (or remove him from it). Rather, we elect representatives to the Electoral College. These electorates vote the overall wishes of the people. When a candidate has enough electorate votes, he has won the election. As a matter of fact, it is possible for the majority of the voters in the United States to vote for one man for President and still have his opponent (who received less popular votes) win. Albeit, the majority would have to be a very small one and would have to come from those smaller states with a proportionately smaller number of electoral votes. Some people have been wondering why we have an Electoral College instead of electing the President by the popular vote.

The answer is simple . . . . Television (or rather, a lack of this medium). When the Electoral College system was first established, the United States was spread far and wide. The same distances separate us today (in statute miles), but television has brought us all closer together. Now, we know on the East Coast what the voters are doing in the West. Commentators may even tell us why they're voting the way they are. We know instantly and accurately. Often, the news commentators predict the winner long before the loser concedes. Sometimes they're wrong, but the vote tabulation rarely is.

But back when the Electoral College system was established, we didn't have television coverage (or radio either). There was simply no practical way to keep tabs on what the entire nation of voters was doing from a central location. The only thing left to do was to throw the burden on each individual state to let the popular votes, which could more easily be kept track of on a state basis, determine who the electorates would vote for and then let state population determine the number of electorate votes each state was permitted. The system was certainly sensible for the times.

Today, however, many persons (some in office, some not) are saying that the Electoral College system is outmoded and obsolete due to the tremendously efficient communications systems that are now available to us here in the United States. Television has played a great part in this area, because almost anyone can learn, first hand, by sight and by sound, just how good our ability to communicate with each other, near and far, really is.

Americans are more opinionated today, again, mainly due to television. We see more events and hear every side of different arguments. We are better able to form opinions and see others whom we admire share those opinions (and those we don't admire usually disagree). No, television shouldn't become an opiate which runs our lives, but rather an information source which helps us enjoy life and stimulates our activities.

Let's discuss another aspect of television, one which introduces some negative aspects. Television supplies information; therefore, it is an active system or media. We receive information; therefore, television viewers are often said to be passive. I guess there is a bit of truth in this statement, although there is a good argument that the passive reception of information often leads to the active application of it. But assuming the former statement is true, even this is changing. Already, a few cable companies with local programming have installed special interfaces on their home hookups which allow the television audience to respond directly to programs by pushing a button. Local elections, referendums, and commonly-shared ideas and problems are discussed on the air. The audience is then asked to respond with a yes or a no by depressing the appropriate button on their home consoles. Now, information is taking a two-way street. The information transmitted by the cable company or television station on the cable is passively received by the home viewers. When they respond, they are sending information back to the studio which assumes the passive role in this cycle of operation. This allows for an accurate measurement of the

4

feelings of a large portion of a local market. This is already being used to advantage by local governments who can now receive information from many more people than the few who will usually show up at weekly or monthly meetings. This helps government to do a better job of serving the people. It also allows the private citizen to take a bigger part in government.

At this point, we have come full circle, from cartoon shows to audience-participation television. Let's begin again by going back to a cartoon show. You may remember this program from your youth if you are in your early thirties; but if not, you will be slightly amazed at the parallel which can be drawn between a child's program and audience-participation TV.

The show referred to was aired in the middle 1950s and was mentioned a bit earlier in this chapter. I believe it may have been partially or wholly developed by Jack Barry, who was the human star. If not, my apologies to the genius who dreamed this idea up. The real star of the show was a real star, the five-pointed variety, and he went by the name of Winky Dink. The cartoon character was a five-pointed star with arms and legs, and he got himself involved in all sorts of cartoon adventures. How can this possibly compare in any way, shape or form with audience-participation TV? Read on!

This would have been just an average cartoon show if not for the basic concept of the entire program. You had to actively participate or the show was completely incomprehensible. This participation also required the expenditure of fifty cents for a Winky Dink Magic Screen, Magic Crayons, and Magic Eraser. The Magic Screen was simply a piece of clear, flexible plastic which would adhere to the television screen without glue. I guess the static created by the picture tube performed an adhesive function. The crayons were standard crayola types, although they were bigger than usual. The Magic Cloth was an ordinary cotton rag.

Now that you're completely confused, let's move on to the system. Shortly before the program started, the Magic Screen was applied to the surface of the picture tube. Jack Barry always told us to rub it flat with the Magic Cloth so that it didn't have any wrinkles. Then, with crayons in hand, we watched for the entrance of the "star Star" Winky Dink. Jack would interview him for a minute or two; then the adventure would start. A special guest was announced and Jack Barry would draw a line on a clear glass panel which lay directly in front of him and was invisible to the TV audience. It seemed as though he were writing in midair. We at home would then trace this line on our magic screens. With the camera locked in place, the first line would disappear and Jack would draw another. A few minutes and many lines later, we kids had traced a perfect kangaroo or other such animal on the Magic Screen. Those home viewers without this attachment saw nothing. Then a mouth would appear at the proper place on the home-drawn character's face (this was produced at the studio through video effects). The animal would begin to talk and the mouth would move with the speech pattern. Again, the audience without a Magic Screen only saw a moving mouth suspended in space with no body around it.

With Jack Barry's help, we drew bridges for Winky Dink to cross raging rivers on and boats for him to sail the ocean. We even drew little rocks which seemed to be hurled across a room as the actual TV scene was quickly moved from right to left or vice versa. We got our first driving experience by sitting behind an automobile's instrument panel which we drew ourselves (ala Jack Barry) and watched the television roadway and scenery go by at a rapid rate. By golly, we participated! Without loyal viewers (and artists) such as myself, ol' Winky Dink would have been a self-destroying supernova in no time at all. We felt as if we were right there with him through all his trials and tribulations, and we were the heroes that saved the day. That is true audience participation and direct involvement.

I don't know why this program finally left the air. It couldn't have been because of lack of an audience. Perhaps the FCC began to take a dim view of programs which could appeal only to those who could afford the fifty cents for the Magic Screen. Of course, long before my Magic Screen arrived, I was cheating the system by using a piece of plastic wrap taped to the TV screen and my coloring book crayons.

It is now time to get to the subject which this book is designed to discuss in detail—satellite reception of television signals at home. Personal Earth stations are getting people involved in television like never before. Participation is required, because it will probably be necessary for you, the owner, to do a bit of research and even install much of the equipment for the station yourself. No, you don't have to be an electronics wizard. Earth stations are for just about anyone, but you will become involved, even if you have a team of technicians come in and install the entire system for you.

Satellite television reception will open up yet another world of entertainment, information, and educational programming which was never before possible. This is certainly the "television age", but we are entering a new phase which will significantly change the lives of many millions of people.

Through the use of a personal Earth station, you will be able to select programs that were never before available and to search out new programs that are offered on a regular basis. You may even become a satellite "hound" who takes pleasure in homing in on exotic foreign orbiters (Russian spy satellites, for example) which you have seen listed in your elevation/azimuth charts.

You, quite possibly for the first time in your life, will be receiving signals directly from outer space, from a tiny sphere or oblong object which lies about 23,000 miles in altitude over the earth's equator. The only major problem you will have will be in deciding which one of the hundreds of programs you want to watch.

It's an exciting world and an even more exciting universe, and there are things which are literally out of this world which you will be depending on and making use of daily. And all of this can be had from the comfort of your easy chair with your personal satellite Earth station snuggled on a small patch of land in your backyard.

Like those early years of black and white television receivers for the home, the owner of a personal Earth station is somewhat of an interesting oddity, one which all the neighbors will want to be friendly with in order to obtain a personal demonstration of this Captain Midnight-like device.

Yes, we are entering a new age of television, and those of us who are at the forefront will be able to take advantage of the increased services satellite television has to offer. As wonderful as satellite TV is now, it can only get better as the state of the art develops. Just like that old Dumont black and white that grew up with the author, your satellite Earth station will grow up in a new age with you.

Chapter 2
Satellite Broadcasting

We are all familiar with the many standard broadcasting services. We can also include such facilities as amateur radio stations, CB stations, and a multitude of other services which use the air waves to transmit and receive information. Any radio system is composed of two basic parts, the transmitting station and the receiving station. A brief discussion of these will better enable us to push onward to satellites and the equipment associated with their uses.

TRANSMITTING STATIONS

Transmitting stations have one purpose: the gathering of information which is discernible to the human auditory and visual systems and then modifying the sending of this information to enable it to be broadcast over generally long distances. Let's take your local radio station, for example. In Fig. 2-1, the announcer speaks into a microphone. The microphone is a transducer. It takes energy from one system and transfers it to another. The human voice is composed of audible energy. The frequency, level, and other aspects of the human voice convey information. This is received at the input to the microphone. This device transfers the information to an electrical system. The vibrations of the human voice cause a diaphragm within the microphone element to vibrate sympathetically. This movement sets up an electrical current flow, which is the equivalent of the human voice.

The output from the microphone is of very low level or amplitude. The next stage in the broadcast chain is to step up this energy, or amplify it. This process is accomplished by audio amplifiers, which accept a small voltage at their inputs and produce a much larger voltage at their outputs. It is important to note here that while the output voltage may be higher, it

8

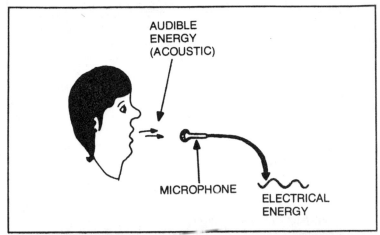

Fig. 2-1. As the announcer speaks into the microphone his voice which, is audible energy, is transferred into electrical energy by vibrations in the microphone element.

changes in such a manner as to directly correspond with the changing levels at the input. For example, if the audio input to an amplifier containing a bit of information which was originally applied to the microphone by a human voice caused the voltage to swing from 1 to 10 volts, then 10 volts is the maximum applied input. Let's assume that the output from the amplifier has a maximum level of 100 volts. Based upon the same amplifier input, the output voltage would swing from 10 volts to 100 volts. The output is much higher, but the relationship of the highest value to the lowest is still 10 to 1. As long as this relationship is maintained, the detected audio information from the output will be the same as that at the input, only much louder. The reader should note that this is a very simplistic explanation of the functioning of an audio amplifier and that there are many other criteria involved in audio reproduction and amplification.

After the voice information has been converted to an electrical system and amplified, it is then passed on to the radio transmitter. If we were to connect a speaker to the output of the amplifier, we would again hear the human voice. The speaker is a transducer which acts like a microphone in reverse. It accepts information from an electrical system and converts it to sound waves. But, we are not ready to demodulate the information contained in the system. We must first use the transmitter to alter the electrical energy from the output of the audio amplifier and place it into a system which will allow this information to be carried long distances. The transmitter effectively takes the low frequency audio information and steps it up to a high frequency. This is done because high frequency waves travel further than low frequency waves. The human voice normally produces frequencies between 300 and 3000 hertz (cycles per second). The broadcast distance of the unaided human voice is very limited (as far as we can shout).

If we were to take this same information and step it up to a high frequency, then the broadcast range would be greatly increased. A good way of demonstrating this principle of high frequencies traveling farther than low frequencies is to use a high school band as an example. The piccolo, which produces the highest frequency output of any band instrument, is easily heard, while the tuba, a low frequency instrument, is not. The level or loudness of the piccolo's output is probably far less than that of the powerful tuba, but since the high frequency travels further, the piccolo stands out. You would have to get very close to the tuba to be able to hear it as well.

The radio transmitter also contains an audio amplifier. In AM transmitters, this amplifier is often called the modulator and actually consists of several audio amplifiers in series. At a 1000-watt station, the final audio amplifier output of the modulator will be around 500 watts. The announcer's voice, which contained less than a watt of power as he was speaking, now packs a walloping 500 watts of power.

The second section of the transmitter produces a high frequency carrier wave. Its frequency is on the order of 1,000,000 hertz (1 megahertz). The carrier serves as a reference. When the amplified audio information reaches the modulator, it is applied directly to the high carrier frequency. Let's assume (to make things simpler) that the announcer is humming a musical note. The frequency of this note is 1,000 hertz (1,000 cycles per second). When this is applied to the carrier, two sidebands are generated. One will be 1,000 cycles per second higher than the carrier frequency of 1,000,000 cycles. The other will be 1,000 cycles less than the carrier frequency. But the entire transmitted wave is no longer in the audio spectrum but in the high frequency range at around 1,000,000 cycles. The upper sideband is broadcast at a frequency of 1,000,000 plus 1,000 cycles. The lower sideband is generated at 1,000,000 cycles minus 1,000 cycles. The upper sideband frequency is then 1,001,000 hertz and the lower sideband is 999,000 hertz.

Now that we have converted from audio frequency to radio frequency, the transmission is broadcast from the radio station's antenna. This is how all basic communication systems operate.

RECEIVING STATIONS

Like transmitting stations, receiving stations have one major purpose: to detect the information from the transmitted wave and convert it to information which can be directly detected by a human being.

Referring to where the previous discussion left off, we now have a radio wave being broadcast from the station which contains audio information (the 1,000 hertz tone). The radio wave spreads out from the transmitter site until it strikes the receiving antenna. The receiving system is shown in Fig. 2-2. Here, the antenna is designed to respond to the frequency of the transmitted wave. It transfers a small portion of this wave to the receiver circuitry. The carrier, having served its purpose, is eliminated and the audio information is retrieved. The purpose of the carrier is to carry the

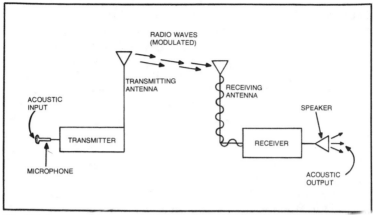

Fig. 2-2. In a receiving system the antenna responds to the frequency of the transmitted wave.

audio information; hence, its name. It also serves as a reference. Without it, the audio frequencies cannot be deciphered and would simply look like a garble of radio frequencies. But remember, the audio information is spaced a specific distance (in frequency) from the carrier. If the input was 1,000 hertz, then the audio information is 1,000 hertz either side of the frequency of the carrier.

The receiver circuits utilize the carrier as a reference. Simply put, the carrier frequency is equal to 0 hertz. The sideband frequencies which are 1,000 hertz either side of zero are then passed on as a 1,000 hertz audio tone. We are still sending information in an electrical manner and it becomes necessary to convert this to sound wave output which can be directly detected by the human ear. A speaker is attached to the electrical system and its output is the same as what was produced at the microphone by the announcer.

This complicated process occurs at the speed of light and is ongoing throughout the entire cycle of transmitting and receiving. An earlier discussion in this chapter stated that if the voltage swing at the output of an audio amplifier is equivalent to the swing at the input as far as percentage is concerned, then the retrieved information will be identical at both points, only at different levels. The same can be said of transmitted information on a radio wave. As long as the frequency swing between the carrier and the sideband is identical to the audio frequency swing, then the information detected at the receiver will be identical to that which was originally put into the transmitter. The frequency of transmitter operation makes no difference whatsoever. Since the audio frequencies of the average human voice cover a range of approximately 300 to 3,000 hertz, the sidebands produced at the output of an AM transmitter will swing from 300 to 3,000 hertz either side of the carrier frequency. This is why you can hear your announcer exactly as he sounds when talking in the studio.

RADIO WAVES

While the frequency of the transmission of radio waves makes no difference to the demodulating process (retrieving the audio information), it is very significant when speaking of how these waves will travel across the surface of the Earth and, indeed, through the Earth's atmosphere and out into space. Most of your AM radio station's transmitted power goes to the ground wave. This means that the transmitted signal travels along the surface of the earth. Mountains, buildings, towers, and other obstacles interfere with the wave and sap power from it. This, along with free space attenuation (the lessening of the amplitude of the signal as it travels farther and farther from the transmit site), means that very, very weak signals are applied to the receiving antenna. Low-powered AM stations of 1,000 watts or less have ranges which are typically less than 20 miles, although this will depend upon flatness of terrain and other criteria. If you are located outside of the broadcast area, the transmitted signals are too weak for good reception or perhaps for any reception at all.

Frequency is the great determining factor in whether or not the output of a transmitter will produce a ground wave. The antenna also will play a part in this. Ground wave communications are limited. This is why there are so many radio stations in the United States.

LINE OF SIGHT BROADCASTS

To partially overcome the limited range of AM broadcast stations whose transmitted power is mostly tied up in ground wave transmissions, there is the FM broadcast band, which lies between 88 and 108 megahertz. This band is at about 100 times the frequency of most AM broadcast stations. Due to the high frequency of operation, FM broadcast band stations normally transmit by line of sight. This means that the signal will be received when the receiving station lies within a direct line of the transmitter. This is shown in Fig. 2-3. Mountains and other obstructions will still create signal losses, but in many areas, reception is more dependable. Line of sight broadcasts are typical of all radio transmissions but especially of those above 50 megahertz.

A major problem comes into the picture when discussing line of sight broadcasts and there's really very little that can be done to overcome it. Figure 2-4 shows two stations on the surface of the Earth, one for transmitting and one for receiving. The Earth is not flat (something we all know) and its curvature is a direct block to line of sight communications when the two stations are separated by enough distance so that their antennas fall behind this curvature. The Earth itself blocks the transmitted signals from the receiver.

Figure 2-5 shows one method of overcoming the curvature problem. A passive reflector has been installed between the transmit and receive stations. This device is within line of sight of the transmit and receive antennas. Now, when the transmitter produces its output, it strikes the

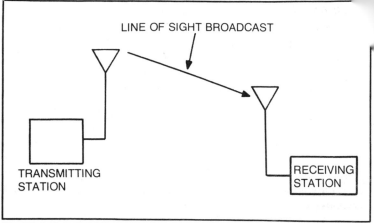

Fig. 2-3. In line of sight broadcasting systems the receiving station must lie within a direct line of the transmitting station.

passive reflector which reflects the signal in the direction of the receiving antenna. Such installations are quite expensive but are often used for microwave communications purposes. The alignment of the reflector with the other two antennas is extremely critical.

Another method uses an active repeater between the two stations. This is shown in Fig. 2-6. When the transmitted wave strikes the repeater, a receiving antenna detects the signal and feeds its output to the input of another transmitter. Its antenna broadcasts to the final receive station located behind the curvature. What this boils down to is another receiver and transmitter between the two sites which are used for communicating information. This is more expensive that a passive reflector, but the signal strength and quality at the final receive location is far better.

SKYWAVE COMMUNICATIONS

The frequencies between the AM broadcast band and the FM broadcast band are utilized by many services. Among them are amateur radio operators, some business band communications, the citizens band network, and a myriad of foreign broadcast stations. Frequencies between 3 and 30 megahertz travel by ground wave and by skywave. Often, a greater portion of broadcast power is wrapped up in the skywave.

Figure 2-7 illustrates skywave communications. This is very similar to the previous example of line of sight broadcasting, where a passive reflector was used. Skywave communications also utilize the passive reflector concept, but here, the reflector is a natural one—the Earth's atmosphere. Referring to Fig. 2-7, signals leave the transmitting antenna and travel skyward. The wave strikes the ionosphere (an ionized layer in the Earth's upper atmosphere) and is reflected back to the Earth. This is often called *skip* because the radio waves skip off the ionosphere. The reflected signal is

13

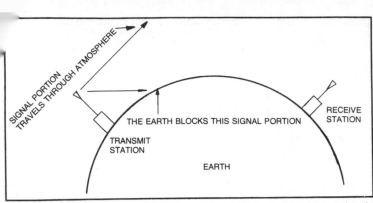

Fig. 2-4. Since the Earth is not flat, the curvature obstructs transmitted signals and prevents them from reaching the intended receiver.

aimed back toward the earth and may touch down at a receiving site thousands of miles away from the transmitting site. The actual distance covered will depend upon the type of antenna used at the transmitter, the frequency of transmission, and the condition of the ionosphere.

Multiple hop transmission may also occur as shown in Fig. 2-8. Here, the reflected wave from the ionosphere strikes the Earth, which also serves as a reflector at these frequencies. The signal leaves the Earth, traveling skyward again to be reflected once more from the ionosphere. This type of skip can take the form of a multi-hop transmission, which will often travel completely around the Earth. I recall several occasions when I have talked from Virginia with a station in California, using the amateur radio band, by multihop transmission. Due to the transmitting conditions and the state of the ionosphere, the signals did not travel across the United States. Rather, they traveled around the world, being received in California after several skip cycles. The amateur operator in California was transmitting by the

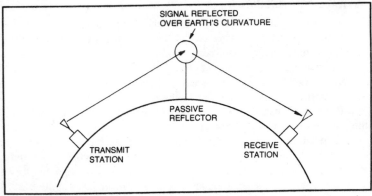

Fig. 2-5. Here, a passive reflector is installed between transmit and receive stations.

14

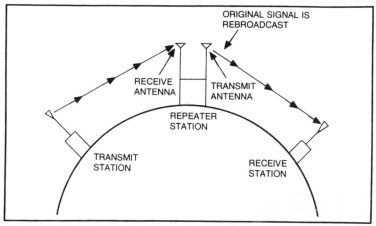

Fig. 2-6. Another solution might be to install an active repeater.

same path. The audio quality of the received signal, along with the reception of foreign amateur operators, were telltale signs that we were communicating by long path DX, a term used to describe these conditions. This is not a tremendously rare occurrence but is exceptional enough to be noted here.

SPACE COMMUNICATIONS

As radio frequencies increase, line of sight becomes the rule rather than the exception. Skip effect begins to disappear. This is due to the makeup of the ionosphere. During skip transmission, only those signals which lie below a specific frequency will be reflected back to earth. Higher frequency

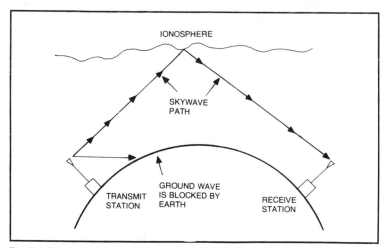

Fig. 2-7. Skywave communications utilize the Earth's atmosphere as a reflector.

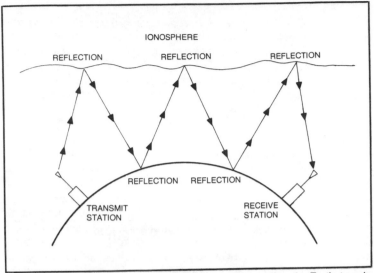

Fig. 2-8. In a multi-hop transmission, the reflected wave strikes the Earth, travels upward to the ionosphere, and then back again.

waves will pass through the ionosphere and out into space. They are lost forever unless reflected back by some means. Meteors can do this, as can the Moon. Another favorite practice of amateur radio operators is called *Moon Bounce.* Here, an ultra-high frequency is transmitted from the Earth, strikes the Moon, and is reflected back to the Earth again. In this case, the Moon is used as a passive reflector.

Orbiting satellites may also serve as passive reflectors, and there is less space loss than when using the Moon simply because the transmitted and reflected distances are far shorter. Figure 2-9 shows a passive satellite system whereby signals from an earthside transmitter travel through the ionosphere and out into space. Here, they strike the surface of a stationary satellite and are reflected back to Earth again. This is still a line of sight system, in that the satellite must be within direct line of sight with the transmitter as well as the receiver. Due to the altitude of the orbiter, many points on the Earth are within direct line of sight. Without the satellite, the signal which is transmitted from the Earth would travel out into space and never return again.

When satellites are used as passive reflectors, many of the same problems associated with earthbound reflectors are incurred and often multiplied. There is a large signal loss between the transmitter and the satellite and an equal loss between the satellite and the receiving station. It requires tremendously high transmitter power in order to assure a strong enough signal back on earth after the reflection process has taken effect.

In an earlier discussion, an earth-based passive reflector was used to enable line of sight communications to take place around the curvature of the Earth. To overcome the problems of signal strength, the passive reflector was replaced with a radio repeater. This device received the transmissions from the sending station and retransmitted them to the receiving station. The same can be done in space communications.

The reflecting satellite just discussed is also called a passive communications satellite. This compares with the earthbound passive reflector. The space communications equivalent of an earth-based repeater is called an active communications satellite. The principle of operation is shown in Fig. 2-10. By studying this, it can be seen that this type of orbiter operates in exactly the same manner as an earth-based repeater.

A high-powered signal is transmitted from the sending station on the Earth. This line of sight microwave broadcast travels through the atmosphere and to the receiver of the active satellite. The output of this receiver is fed directly to a transmitter within the satellite and a new signal which contains the same information as the old one is transmitted back to the Earth. Even when very high amounts of power were used by the Earth transmitter in a passive satellite system, the signals that were received back on Earth were extremely weak. The same high amounts of power must still be used with an active communications satellite system, but since the signal is retransmitted out in space within the satellite proper, the received transmissions back on Earth are far stronger.

The active satellite system is the one we are interested in as far as television receive-only Earth stations are concerned. The original broadcasts are transmitted from very high-powered Earth stations whose antennas are aimed directly at the satellite. These stations transmit on a frequency of around six gigahertz. Out in space, the active satellite is equipped with transponders. A transponder is simply a transmitter and receiver which are connected to one another. The detected information from the

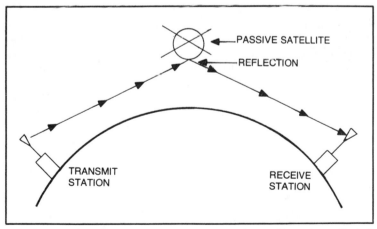

Fig. 2-9. In a passive satellite system, signals strike the satellite and are reflected back to Earth again.

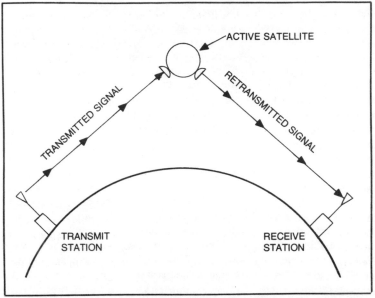

Fig. 2-10. An active communications satellite is equipped with a transmitter, which serves to amplify the transmission before sending it back to Earth.

receiver is fed to the input of the transmitter and beamed back to Earth. One satellite may have a few transponders or possibly even twenty or more. When receiving satellite TV at home, each transponder serves as a separate television channel which can be selected in much the same manner as is done today with your present set.

Satellites are usually physically small devices and space must be conserved wherever possible. For this reason, a satellite with many different channels (transponders) may have only two antennas, one for transmit and one for receive. Each channel shares these antennas by using them for small fractions of a second. To provide an easily understood example, let's assume that a satellite has two channels which must share the same antennas. Channel A may use the antennas for 100 milliseconds and then be switched off for an equal amount of time while channel B uses the system for 100 milliseconds. When channel B is switched off, channel A uses the antennas again. Channels A and B will be switched on and off many times during the blink of an eye, but you could never tell this by watching a received picture at your Earth station because of the speed with which the switching occurs. If you were watching channel A, you would never know when channel B was using the same antennas to transmit on another frequency which you were not receiving. Using standard television sets and earthside broadcast stations, the video picture you receive at home in one second is actually about thirty different pictures. Without getting too technical, the television picture can be likened to the easel of a very fast artist. Every picture that is painted is stationary. It has no movement whatsoever. But when the next picture is painted, the figures have changed slightly and the change is even greater when the next picture is painted. When thirty different pictures are painted per second (as is the case on your home television), things happen so fast that we do not sense the time intervals between picture tube sweeps. Therefore, we view the television picture as a continuous sequence of events which happen in periods of seconds rather than milliseconds. Human beings are incapable of sensing or responding to events which occur in small fractions of a second. Because of this, we can take a look at a television screen where many thousands of things are taking place in these small fractions of a second and automatically see what has occurred by averaging these events over time periods of seconds.

To clear this up a bit more, refer to Fig. 2-11 which shows a ball being thrown. This picture shows the position of the ball at many points along its trajectory. Any one of these points could represent a fraction of a second of the time it took for the ball to leave the pitcher's hand and strike its target. We humans do not perceive things in this manner, however. In viewing the pitch, we would average the events which took place during this throw over a time span of one or two seconds. We would see the total event rather than the tiny segments which are shown in the drawing. Computers and other electronic circuits can be said to think in millionths of a second. They can receive information, perhaps a thousand times faster than us. We could say that computers are very fast, but computers could say we humans are very slow. If a computer had self-awareness, it would see human beings as extremely slow-moving objects, because we think and act in a completely different realm of time. Think of it. If it takes you one minute to tie your shoes, it would take a computer (if it were capable of this type of motion) one-thousandth of a minute. On the other hand, if you were the computer, one-thousandth of a minute would seem to you to be about as long as a full minute is to normal human beings. Therefore, after you tied your computer shoes, you would *seem* to wait a full one thousand minutes for your human counterparts to tie his.

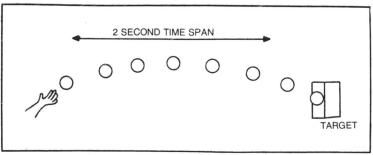

Fig. 2-11. Although in this drawing the ball is seen at all the positions it travels, humans perceive this in a single motion rather than as fractional segments.

In any event, the on/off nature of satellite transponders is totally unnoticeable by human beings who cannot think in small fractions of a second. While our previous example used a satellite with only two channels, most have more than this and each shares the receive/transmit antennas for correspondingly smaller periods of time.

As was stated earlier, most television stations which use satellites for their broadcasts use very high-powered transmitters on the ground. They transmit at a frequency of approximately six gigahertz, but these signals do not come back to earth at the same frequency. Referring to Fig. 2-12, the six gigahertz transmission leaves the earth and travels to the satellite. The orbiter's receiver is designed to detect a six gigahertz signal. It pulls the audio and video information from the transmission and then feeds it directly to the satellite transmitter, which has an output at a frequency of about four gigahertz. This is the frequency which the earth-based television receive station is set up to detect. A further explanation of frequencies will be found in future chapters.

Since we know that the satellite signal is originally transmitted on the Earth, travels into space, is retransmitted by the satellite at a different frequency and is finally received back on Earth again, it can be said that the signal that is ultimately received by an Earth station is a function of:

- The signal transmitted to the satellite at six gigahertz.
- Signal processing in the satellite.
- Signal transmitted at four gigahertz from the satellite.
- Directivity of the satellite antennas gain.
- Path loss.
- Gain of the receiving antenna.
- Noise temperature of the antenna.
- Low noise amplifier noise temperature and gain.
- Cable loss to the receiver.
- Receiver noise figure.

All of these terms may not be familiar to you and later chapters will clear up many of the questions. This list is presented to show the many factors which determine how well a signal is received by your personal Earth station. Many of these functions need not be dealt with directly, as they are taken into account in computer reports which can be ordered inexpensively to describe receiving conditions in your area.

There are many different satellites in orbit around the Earth. Entertainment satellites such as the ones used by personal Earth station enthusiasts are in special orbits which make them stationary in relation to the earth. To us, these satellites are always in the same location. The Appendices provides a listing of present domestic satellites, along with their longitudinal locations and additional information on the various programs they carry.

It is from these satellites that a whole new world of television enjoyment emanates. Sure, all of the signals are originally transmitted here on

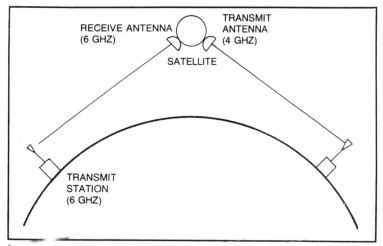

Fig. 2-12. Signals which leave the Earth at one frequency do not return at the same frequency.

Earth, but with a satellite deep in space, these transmissions can be received by so many more persons. The wonderful thing about receiving signals from satellites is found in the fact that we tend to pay very little attention to these multi-million dollar orbiters. We know they're up there and we know their positions in order to properly aim our antennas; but other than this, the Earth station equipment is the part of this complex system which gets the most attention. The satellites are reliable, always present, and simply do not require the attention that our personal Earth stations require. Billions upon billions of dollars of research have gone into making our satellite program as dependable and useful as it is, and while most of the reader's interest will be taken up in the building of his own personal Earth station and not concentrated on the satellite itself, an understanding of what has been done over the years to enable that satellite to be overhead is important.

A new era for all mankind opened in 1961 when Major Yuri Gagarin of the Soviet Union was rocketed into space. He was followed shortly thereafter by Commander Alan B. Shepard, Jr. of the United States. Since that time, men have been launched into Earth orbit so many times as to barely warrant a news story. Americans have been to the Moon and space communication equipment was more than adequate to send back beautiful color pictures for all the world to see. None of this would have been possible had it not been for the many years of experimentation and testing of satellites that preceded manned flights. The first satellite was Sputnik, which was launched into orbit by the Soviets in 1957. This tiny device did little more than beep a weak signal back to Earth. It weighed 184 lbs. and was quickly followed by Sputnik II, which weighed 6,610 lbs. The first Sputnik was programmed to record temperature and pressure. The second orbiter read solar and cosmic radiations, along with temperature and pressure. The Sputnik was followed by the American satellite Explorer I. This satellite weighed 31 lbs and was used to measure temperature, cosmic rays, and micrometeorites. All of these satellites and the many more to follow had one thing in common: they sent radio signals and television signals back to Earth.

It was long known that rockets could be tracked by radio. This simply involved the placing of a radio transmitter with a very stable output in the nose cone of the rocket. Receiving stations on the Earth could detect motion of the rocket from the condition of the radio signal. There are two very different purposes for radio transmissions in rockets and satellites: tracking and telemetry. Tracking simply tells you where the orbiter is in relationship to the Earth or other objects. Telemetry is a laboratory measurement, only very rarely concerned with position finding. The telemetry transmitter receives its input from instruments carried in the satellite. The detected results at the Earth receiving station are converted back to the original information.

It can be said that satellites used for television reception are a form of telemetry. The sensed information from space is the incoming transmissions from the Earth sending station. The signal is processed within the satellite and then transmitted back to Earth again, where it is detected by the personal Earth receiving station. The information is pulled from the signal and converted to audio and video output which feeds the television set. This is the same manner in which we received color television pictures from the Moon during the last lunar landing, except the transmitted signals originated on the Moon instead of on the Earth.

As satellite experimentation continued, the equipment became much more dependable and highly efficient. In the early 1960s, the communications satellite Telstar was launched into orbit and it suddenly became possible for us to receive pictures directly from the other side of the world. Before this time, television programs were recorded at distant points on the globe and played back in the United States at a later time. Most of us today take for granted coverage of the Olympics, for instance, which we receive via satellite. It seems like no great thing that we can see the swimming competition here in the United States at the same time this event is occurring in Germany. Today, any point on the globe can serve as a temporary television studio because of the satellites orbiting the Earth. Such portable communications would have been totally unknown just twenty years ago when television broadcasts had an average range of less than 150 miles. Today, the same broadcasts can circle the earth.

I saw the first telecast via Telstar when I was a boy. The picture quality was not tremendously good, nor was the audio. Back in those days, you could still tell the difference between a standard broadcast and one that used satellites—today, you cannot. Space communications technology has made vast strides since the 1960s and a picture received from 23,000 miles out in space is just as clear as one received from your local television station across town. This, of course, assumes a properly installed and operating satellite television Earth station.

This is truly a wonderful aspect of a technological age. Satellites open up many possibilities for all types of people. Satellites have been designed and built by amateur radio operators working in conjunction with space communications technicians and then blasted into orbit as part of a satellite package by the U.S. government. Amateur operators can now communicate via satellites. Telephone conversations to the other side of the world no longer use expensive cables lying on the ocean floor. They use satellites. Anything which can be sent by hard wire can also be sent by means of satellites. And this is just the beginning. Think of the strides which have been made in space communications over the past twenty years. You will see that the last five years represent an even faster technological growth which is bound to accelerate in the coming years. We are bound to see more and more satellite services being opened.

One of the recently accomplished strides was the successful launching, orbiting and return of the Space Shuttle Columbia. Satellites have limited effective life spans of anywhere from three to eight years. After this time, their orbits begin to decay until they finally burn up in the Earth's atmosphere. But the Space Shuttle now gives us the capability of going out into space, picking the aging satellites up and servicing them inside the ship. After everything has been checked, the satellite is placed back into orbit again. This corresponds to your television repairman coming to your home to make repairs on your set. You don't throw away your television set when it breaks, so why do it with satellites which cost millions upon millions of dollars? The Space Shuttle also will open up many thousands of possibilities for satellite users. This ship is quite capable of placing many satellites in orbit during a single trip. Its massive cargo hold is adequate for carrying satellites to and from the Earth. Instead of launching one or two satellites with huge rockets which are used only one time, the Space Shuttle is used again and again and will carry a larger payload. This is bound to decrease the costs associated with building communications satellites and placing them in orbit. The Shuttle should also be able to assure the large corporations which own these satellites that they won't be lost after five or six years. From the space communications standpoint, we are entering the most exciting era of all time. Those of us interested in setting up our own satellite television Earth station are on the edge of a new frontier.

SUMMARY

Satellite broadcasting has opened up limitless opportunities for the average person. We have been realizing the advantages of space technology for many years. Previously, satellites were used strictly for government, military, and scientific purposes. In the early 1960s Telstar brought satellite communications into every home. But even then, and for many years thereafter, the average person had little control over what he could receive. Now, with personal satellite television Earth stations being offered at affordable prices, a tremendous variety of programs can now be received from space. This field can only continue to grow. New programs and programming services are being offered on a regular basis. All the average individual has to do is properly equip his receiving site with the electronic devices which will enable him to tap this bountiful resource. Throughout this book, we will be exploring the many methods being used by average persons to effectively use what is being made available to all of us. As the state of the art advances, equipment prices should drop and programmers will become more and more competitive by offering information, entertainment, and a myriad of other services which will be aimed directly at a mass of individuals rather than to a few scientific, government, and military installations.

As additional satellites are placed in orbit, the personal Earth station will become more and more valuable. It has been predicted that within this decade, the satellite television service will become more and more of a replacement for conventional television broadcasts. The versatility of space communications is far superior to Earth-only transmissions since more people are able to take advantage of the many broadcasts. Development in areas such as this tend to become aligned directly with user needs and demands. As more individuals assemble their own Earth receiving stations, the industry will respond. Already, there are plans for more satellites and more programs. You, the consumer, are the one who will reap the benefits of these achievements.

Chapter 3
The Personal
Earth Station

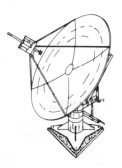

You are probably in a bit of a daze as to exactly what is involved in receiving entertainment channels which originate on Earth, travel over 22,000 miles into space, and then back again. Surely, you must be thinking, satellite reception must involve extremely complex electronic circuits and probably requires an engineer or two to be properly set up and adjusted. This is only half true. Yes, the reception of signals from stationary satellites does involve highly complex electronic circuits, but no, a personal Earth station docs not require an engineering degree to be set up, aligned, and maintained correctly.

Figure 3-1 will be helpful in explaining a basic Earth station designed to receive satellite transmissions and convert them into audio and video signals which are applied to the television receiver by means of standard coaxial cable. While this is a simplistic diagram, it takes in the units which you will purchase or build and assemble yourself by interconnecting them in the order shown.

Let's start first at the front of the receiving end in a personal Earth terminal. As shown in Fig. 3-1, this is the antenna system. The antenna picks up the transmitted signal from the satellite in exactly the same manner as your present television antenna picks up signals from the television station. True, this antenna will look different from your TV antenna. This is due to the fact that a completely different set of frequencies is being used. Figure 3-2 shows a typical TVRO (television receive-only) antenna system. We refer to this as an antenna system because several elements are included. The actual sampling portion of the antenna is also called the feed horn and is the smallest portion of the entire structure. The feed horn is where the actual signal is detected. This, in turn, is relayed on to the electronic equipment which will convert the satellite signal into a

25

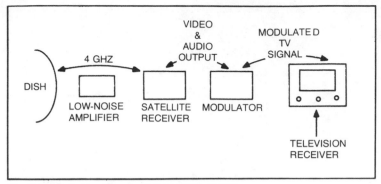

Fig. 3-1. Block diagram of a typical Earth station.

signal which is usable by the television receiver. What's that massive dish-shaped structure for then? This is the passive part of the antenna system, and it is the portion which the satellite transmissions first come in contact with. The large dish is able to sample a much larger portion of the transmitted signal than is the tiny feed horn. The construction of the dish allows all of the satellite signals which come in contact with it to be focused at the feed horn input. This allows the TVRO system to pick up a great deal more of the signal than would be the case if the feed horn were used without the dish.

The construction of the dish is rather critical. It must be aligned in such a manner as to reflect all incoming signals to a very fine point a short distance away from the dish center. It is at this exact point that the feed horn is mounted. The point where all incoming signals are focused by the dish is called the focal point, a term often used in astronomy and when dealing with lenses in general. For this reason, microwave antennas of the type being described are often called microwave lenses. Figure 3-3 shows a pictorial drawing of the basic operation of the dish and feed horn portions of the antennas. A more thorough explanation of microwave antenna operations will appear later on.

It can be seen from this drawing that the dish acts as a sort of funnel, in that every signal which strikes it is brought to a small point. In this manner, much more signal is able to be detected at the feed horn. We can make a further comparison by placing two empty soda bottles outdoors on a rainy day. One is fitted with a funnel, while the other is left as is. After a few minutes, it will be obvious that the one using the funnel fills much faster. The same amount of rain is being applied to both bottles, but the funnel allows for more collection. If you think of the unadorned soda bottle as a feed horn and the one with the funnel attached as the same feed horn with a dish, then thinking of the rain as satellite transmissions, it can be seen that the latter one is able to sample more of the downpour. Every drop of rain which hits the enlarged surface of the funnel is channeled to a focal point at its narrow end. This should provide a good idea of just how a dish antenna

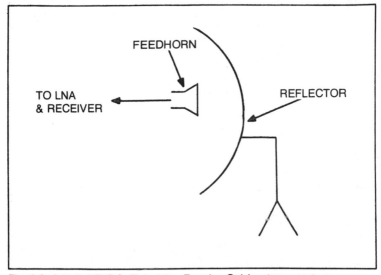

Fig. 3-2. A typical TVRO (Television Receive-Only) antenna system.

operates. The sole purpose of a TVRO antenna is to collect as much radiated energy as possible.

Carrying this last statement further, it is easy to understand that the bigger the dish, the more signal that is applied to the receptor or feed horn. As a general rule of thumb, antennas used for satellite TV reception should be at least ten feet in diameter, but larger diameters are to be preferred, since they will supply more signal strength to the feed horn and thus to the receiving equipment. Some manufacturers make dish antennas in the ten foot range which can later be converted with a special kit to a diameter of twelve to fourteen feet or more.

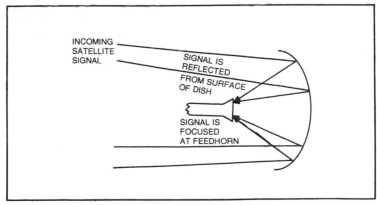

Fig. 3-3. This drawing illustrates the operation of the dish and feed horn.

Once we have gotten as much signal as possible to the feed horn, another set of events are put into motion. We must now provide a means for the signal to travel on to the receiver. But before doing this, we must take another fact into account. While the properly designed dish antenna does an excellent job of concentrating satellite signals, we still do not have enough signal strength to provide adequate input to even the finest solid-state receiver. It is necessary to amplify the voltage which is induced at the feed horn by the incoming transmissions. Here, almost every TVRO Earth station uses an *LNA* (low noise amplifier). To put this in terms which are more familiar, the LNA is a preamplifier. By definition, this device is an amplifier circuit which responds to very low inputs and faithfully reproduces them at its output. The strength of the output signal is not great but it is adequate to drive a standard receiver. Preamplifiers are used in many PA systems in order to bring the level of the very low signal voltage from the microphone up to a point where it can be easily used by the main or power amplifier. All modern radios and television receivers incorporate preamplifiers in their designs. These act directly upon the signals which are supplied by the antenna lead-in cable.

While we are making comparisons here with standard broadcasting services which most of us are at least partially familiar with, specific problems develop at the frequencies used for satellite transmission. A commercial AM broadcast station transmits at a frequency of between 0.6 and 1.6 megahertz (1 million cycles per second). Commercial FM stations transmit their signals between 88 and 108 megahertz. Satellite transmissions for purposes of home television reception are transmitted within a frequency range of 3.7 to 4.2 thousand megahertz. This latter term is shortened to gigahertz, meaning a thousand megahertz. The abbreviation for gigahertz is GHz. Frequencies in this range are referred to as microwave transmissions. The previous two frequency ranges lie within the shortwave spectrum.

At microwave frequencies, physical materials tend to behave differently than they do in the shortwave spectrum. A short length of copper wire which serves as a simple conductor of electricity in an AM radio receiver might look like a very large coil or inductor to incoming microwave signals. Most persons who have experimented with circuits are far more familiar with low frequency operation that those used with TVRO Earth stations. The same theoretical rules apply, but it's a completely different world when it comes to actual practice.

This brief explanation of microwave frequencies is necessary in order to properly explain the attachment of the LNA and, for that matter, the rest of the receiving equipment as well. Whereas most communications receivers incorporate their preamplifiers within the major part of the circuitry, this is not done at microwave frequencies. Coaxial cable which is very efficient in the shortwave spectrum for relaying signals between the antenna and receiver is very inefficient at microwave frequencies. A great deal of signal loss will be introduced into the system by channeling these

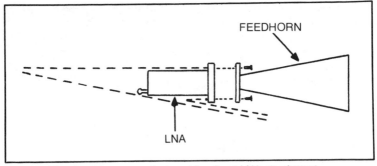

Fig. 3-4. Most feed horns are designed for direct LNA attachment.

frequencies through even the best types of coax. The LNA is designed to operate from very low signal levels which serve as its input, but the signal loss between the feed horn and this amplifier might be enough to render it completely useless, especially if a sizeable length of cable is involved.

This problem is easily overcome by inserting the LNA directly at the feed horn instead of at the receiver. When a direct attachment is made between the feed horn and the LNA, your signal loss is cut to a minimum. Most TVRO antennas are designed so that the LNA is attached directly. This is shown in Fig. 3-4. The LNA, then, becomes a physical part of the antenna structure.

The output of the LNA contains the same information as the output of the feed horn. The signal strength, however, is greatly amplified and can better withstand the rigors of traveling over an inefficient cable to the receiver. This cable serves two purposes, the first being to carry the microwave signals to the receiver, and the second to supply power to the LNA circuitry. The power supply is located within the receiver unit. The two signals intermix quite well due to the fact that one is dc in nature, while the other is at the opposite end of the frequency spectrum.

Now, we have sampled the satellite transmissions, collected them in the dish, focused them to the feed horn, and amplified them with the LNA. The coaxial cable attached to the output of the LNA channels the signal to the next major stage of the TVRO Earth Station.

THE SATELLITE RECEIVER

The end of the cable opposite the LNA is attached directly to the input connector of the satellite receiver. This device is shown in Fig. 3-5 and corresponds to the standard radio or television receiver you presently have in your home. We cannot connect the LNA output directly to your color TV console, because it is not designed to operate at microwave frequencies. The satellite receiver is. While it contains no picture tube or speaker system, this device takes the incoming signal and converts it into video and audio channels. The receiver is a detector which is able to grasp the video and audio information from the microwave signal. This is called demodula-

OUACHITA TECHNICAL COLLEGE

Fig. 3-5. A satellite receiver is similar in operation to that of a radio or television set.

tion. Just like your home television receiver, the satellite receiver will respond to many different channels and is switched accordingly. One satellite may transmit ten or more different channels, so it is necessary to be able to choose the one you want. This is done in the same manner as with your home television—by turning a dial.

The detected audio and video signals could be applied directly to a television receiver, but it would be necessary to bypass the tuner and other demodulation circuits which it contains for the direct conversion of standard TV signals. This is not practical, as it would require circuit modifications to your television receiver. Instead, the video and audio outputs are connected to another portion of the TVRO Earth station.

The video and audio information which are output by the receiver are in detected form. It is no longer traveling on a radio wave. Your standard home television receiver is intended to accept a radio frequency at its input and then to demodulate it and pass it on to the video and audio circuitry. Since the satellite receiver has already performed the demodulation processes, it is necessary to put this information on a radio wave within the frequency range at which the home television receiver is designed to operate. This is accomplished using a modulator, which is sometimes called a downverter. This latter terminology is not exactly accurate, as the modulator actually steps up the frequency of the detected satellite transmission to standard television frequencies.

The modulator is actually a transmitting system. It produces a radio frequency output or carrier wave at standard vhf or uhf TV frequencies. The demodulated output from the satellite receiver serves as an input to the modulator. Now, the radio frequency output of this latter unit is modulated with the demodulated output of the satellite receiver. Figure 3-6 explains this a little more clearly. Most of us are familiar with simple modulators. These are often sold as wireless microphones and are devices which transmit a low-powered carrier frequency which falls within the range of commercial AM or FM broadcast receivers. The input of this device is usually a microphone, and when noises are detected, the carrier is modulated. This is basically how the TVRO modulator works, although instead of a microphone, the circuit gets its input from the audio and video outputs of the satellite receiver.

As will be noted in other chapters of this book, some manufacturers incorporate the modulator portion of a TVRO system within the satellite receiver itself. In these instances, the output from the LNA is connected to the receiver in the usual manner, but the output of the receiver may be connected directly to the home television set. Combination receiver/modulators offer the advantage of less component and wiring complexity in regard to connecting the various major components of a TVRO system. There are less interconnecting cables as well and signal loss due to longer travel paths and connector mismatches may be decreased.

After leaving the modulator, the transmission which originally came from a satellite in space at microwave frequencies now looks (to the television receiver) like any ordinary television signal which is broadcast here on Earth. The receiver happily takes this input signal which is attached by a standard length of RG/59 70-ohm coaxial cable. It is necessary to tune the home television receiver to the channel which corresponds to the frequency output of the modulator. Often, channels 2 and 3 are used as outputs in vhf modulators. Uhf modulators may be adjustable over the entire television uhf spectrum.

That's basically all there is to a TVRO Earth station. While appearing to be physically different from an earth-only television transmission/reception operation, the actual working system is very much the same but operates at higher frequencies. In order to emphasize this point, let's look at a standard television station/television receiver system. Referring to Fig. 3-7, a video image and sound are detected in the studio of the television station and serve to modulate the video and audio carriers in the transmitter. This information is broadcast at vhf or uhf frequencies through an appropriate transmitting antenna. When the signal leaves the antenna, this

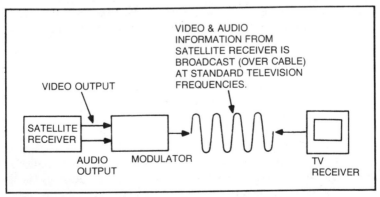

Fig. 3-6. A TVRO modulator is actually a transmitting system.

corresponds to a signal leaving the transponder of a satellite. Now, the earth-based transmission travels through the atmosphere until it is detected by a television antenna. The satellite signal does the same thing, except it must travel through open space and the Earth's atmosphere and in a more direct path to the area of reception.

When the earth-based television signal crosses the home antenna, an electrical voltage is induced which travels down the transmission line cable to the back of the television receiver. Here, the receiver preamplifier increases the strength of the tiny voltage which was induced in the antenna. This is passed on to audio and video demodulator circuits, which extract the original information that was picked up by cameras and microphones at the television studio.

The satellite system compares with this, in that the signal from the orbiter is detected by the dish antenna. It is not passed directly to the television receiver but must first be increased in amplitude by the LNA. It's interesting to note that many fringe area standard Earth television installations for normal reception may use preamplifiers attached to the yagi antennas which are most often used for conventional television reception. You have probably seen these offered by many different manufacturers. Television preamplifiers are the vhf and uhf equivalents of the LNAs which operate at microwave frequencies.

After leaving the LNA, the satellite signal passes through a transmission line cable to the satellite receiver. This exactly parallels what happens in a conventional television-receive system. The satellite receiver also has a preamplifier which boosts the incoming signal to a level which may be used by the video and audio demodulation circuitry. The only difference between the satellite receiver and your home television receiver is that the former does not have a means of making the detected audio and video information directly usable by human beings. In other words, the satellite receiver does not have a speaker and a picture tube.

The reader may be surprised to learn that at this point in the satellite system, a repeat of a complete conventional earth-based television station occurs. The output from the satellite receiver can be thought of as the video and audio information which is picked up on the camera and microphone in the television studio. This information is applied to the transmitter section of the modulator in the TVRO system. Only the standard antenna is eliminated here and replaced with a cable that goes directly to the back of the television set. As was mentioned earlier, the signal received at your set is exactly like the ones which are transmitted from the studio.

A SUMMARY OF THE BASIC TVRO EARTH STATION

While TVRO Earth stations are made up of some fairly sophisticated electronic devices, the actual installation of one as well as the operation is fairly simple and can be accomplished by almost anyone. True, many of the components seem unconventional, but this is only because most persons are accustomed to dealing with the reception of television signals which are

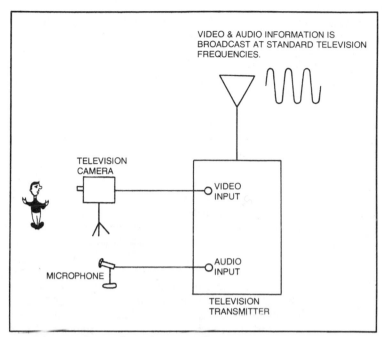

VIDEO & AUDIO INFORMATION IS BROADCAST AT STANDARD TELEVISION FREQUENCIES.

TELEVISION CAMERA

VIDEO INPUT

MICROPHONE

AUDIO INPUT

TELEVISION TRANSMITTER

Fig. 3-7. In a standard TV station / receiver system, video and sound are detected in the studio and broadcast by means of a transmitter.

transmitted at vhf or uhf frequencies and not in the microwave region. Some of the antennas and components used for vhf and uhf frequencies would also seem quite unconventional to persons accustomed to dealing only with dc or high frequency circuits. The high frequency spectrum falls between 3 and 30 megahertz. Vhf frequencies are from 30 to 300 megahertz, while the uhf spectrum begins at 300 megahertz. As frequency increases, the physical dimensions and shapes of components such as inductors, capacitors, and antennas can change drastically. From a theoretical standpoint, an antenna, for example, works in exactly the same manner at 3 megahertz as it does at 3 gigahertz. Why don't you see any dish antennas designed for operation at 3 megahertz? This is due mainly to physical limitations. As a general rule of thumb, when frequencies increase, the physical size of antennas and some other components decrease. A quarter wavelength at 3 megahertz is 82 feet. Therefore, a quarter wave antenna for this frequency would require an element 82 feet long. But at microwave frequencies, the size decreases drastically. At 3 gigahertz, a quarter wavelength antenna would be 0.984 inches in length. Therefore, a 10-foot dish antenna at 3 gigahertz represents a structure which is many, many wavelengths in diameter. But at 3 megahertz, the same dish would be equal to less than .04 wavelengths. To get the equivalent gain at 3 megahertz that we do at 3 gigahertz would require a dish antenna that would span nearly two miles. A wavelength is the physical distance in meters that a radio signal will travel during one cycle. At 1 megahertz, a cycle is generated one million times per second. So the wavelength at 1 megahertz is the physical distance the radio wave will travel in one-millionth of a second. Radio waves travel at the speed of light regardless of frequency. Now, at one gigahertz, a single cycle is generated in one-billionth of a second. Therefore, it can be seen that since all radio waves travel at the speed of light, a single cycle will travel less distance in a billionth of a second than it will in a millionth of a second, just like your family car traveling at sixty miles per hour will not go as far in one minute as it will in two minutes.

Back to the previous discussion, a ten foot dish which offers X amount of gain at three gigahertz would offer the same amount of gain divided by one thousand at three megahertz. You would have to increase the size of the dish by a thousand times to equal the same amount of gain at the lower frequency. In other words, a dish antenna at three megahertz which would equal the gain of a ten foot antenna at three gigahertz would have a diameter of merely 10,000 feet.

The purpose of this comparison is to show that physical limitations in size dictate the use of different types of antennas than the ones we are accustomed to seeing. From an antenna standpoint alone, we can easily build models with extremely high gains which fill a relatively small space. Unfortunately, this attribute is offset by the complexities involved in building efficient circuits to handle these microwave frequencies. As wavelength becomes smaller and smaller, physical circuit components such as resistors, capacitors, and even conductors begin to assume the qualities of other components. As will be mentioned several times in this book, a single short length of copper conductor can appear as a very sizeable inductance at microwave frequencies. Most transistors are completely useless due to their physical construction and will not be efficient much above the vhf or uhf range. The construction of inductors is highly critical. Those of you who have had the opportunity to wind coils for high frequency circuits know that a slight variation in spacing the turns will have little effect. At microwave frequencies, circular coils are not used for inductors, but the equivalent of a minor spacing error would be disastrous in this upper spectrum. At three megahertz, a difference of a couple of inches in the length of an antenna which is supposed to be 82 feet long will not have a great effect on its operation, since three inches is an extremely small percentage of the total length. But at microwave frequencies where the same antenna would be measured in fractions of an inch, a minute error could render the antenna useless. At microwave frequencies, we are dealing with extremely small physical sizes in frequency-determining sections of receivers, transmitters, and LNAs. It is difficult to achieve micrometer-like accuracy in building many circuits. This is one reason why some types of microwave equipment cost so much and also why home building of microwave circuits is not nearly as applicable as it is in the lower frequency ranges.

So that you're not scared to death, it should be understood that rarely are dimensions of one quarter wavelength used in the microwave spectrum. Most receiving and transmitting components use frequency-determining elements of dimensions one wavelength or more. Using our previous example of a quarter wavelength antenna, theory in this area will indicate that a ¾ wave antenna or any radiator with an old multiple of a quarter wavelength will act very much like the original quarter-wave design. For practical purposes, it would behoove design engineers to use an antenna element 21 quarter wavelengths long, as an example, than one of one-quarter wavelength. They would both behave in the same basic manner, but the larger one would be easier to design to tight tolerances because of its length and the lower percentage of error that would result. By this we mean that an error of one one-hundredth of an inch would be a lower percentage of the larger antenna than it would be if we used a tiny quarter-wave design.

Fortunately, anyone interested in their own TVRO Earth station will not have to worry too much about the fine details of the principles we have lightly touched on here. This discussion was offered in order to show the reader that he is dealing with a whole new ball game at microwave frequencies. It is absolutely essential to bear in mind that at these frequencies, losses can develop by some of the tiniest errors that would be completely unnoticeable in other types of systems.

Since the final output of a TVRO Earth station will be at standard television frequencies, it is desirable to get from the microwave range to the vhf spectrum in as short a space as possible. A TVRO system operates at microwave frequencies from the antenna to the LNA and finally, to the satellite receiver. From this point on, we are back in the realm of longer wavelengths.

We know that it is easier to incur greater signal losses within the TVRO system at microwave frequencies than at vhf frequencies. Therefore, special attention must be given to the connections between the antenna, the LNA, and the satellite receiver. After this point, the lower vhf frequency to the television receiver prevents few loss difficulties. Short interconnecting cable lengths must always be used between the LNA and the receiver. This is where a great deal of loss can enter a system. Some manufacturers of satellite equipment offer receivers which must be located at the television receiver in order to change frequency. This means that the transmission line between the LNA and the receiver must travel a longer distance than if the receiver were located right at the dish antenna. This may work well in areas where strong satellite signals are received and where a special low-loss transmission line is used between the LNA and the receiver. But in marginal areas, the loss incurred in the cable may result in poor reception. Figure 3-8 shows the installation under discussion.

Most manufacturers also offer the same receivers as discussed above but with a remote tuning option. Figure 3-9 shows a satellite receiver which must be mounted near the television receiver, while Fig. 3-10 shows the

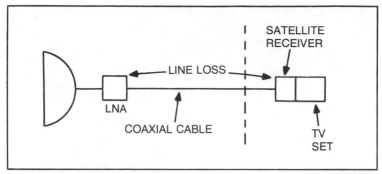

Fig. 3-8. Block diagram of a TVRO Earth station showing a receiver mounted near the television set.

same receiver with a remote control option. Both contain nearly identical circuits, but the latter one allows the microwave portion of the receiver to be mounted at the antenna site while a separate control box is used to remotely tune to different channels. The latter system is more efficient, as it incurs less losses in signal strength. Using this system, the receiver portion would be mounted in back of the dish antenna, possibly in a waterproof enclosure, although most are enclosed for protection from the elements. Some manufacturers of dish antennas offer a special platform at the mounting base for the location of these receivers.

Once the receiver is installed, its output may be connected directly to the modulator or if one is included in the main circuit, the output line is connected directly to the television set. The remote control unit sets atop the television and is connected by a multiconductor cable. When the frequency knob is turned, the remote receiver automatically switches frequency. Alternately, the output of the remote receiver may be connected by coaxial cable to the modulator, which also sets atop the television receiver. Remote controlled receivers are mandatory whenever the television receiver is located a great distance from the antenna site. In some cases, the antenna may be mounted in a backyard just a few feet away from the receiver. Here, you may get away with having the entire receiver directly at the television set without having to go to the expense of obtaining the remote control option. In other installations, it may be necessary to locate the antenna at a site some distance from the home which offers a better window on the satellite. This is where a remote control receiver must be used. Some antenna manufacturers even offer remote antenna adjustment controls which will rotate the dish in order to allow for the reception of signals from several different satellites.

While this discussion has stressed the importance of preventing losses in the microwave sections of the TVRO Earth station and indicated that there was not as much concern about losses in the vhf transmission portion, losses can occur at vhf frequencies as well. It is simpler to prevent losses at these lower frequencies, but it is quite important to pay attention to this

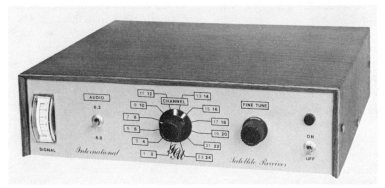

Fig. 3-9. In this type of installation, the satellite receiver must be located at the television receiver.

area with as close an inspection as is mandated for the microwave section. The Earth station owner must make certain that all cables throughout the system are of high quality and in good repair. The manufacturers of the various pieces of equipment for Earth stations usually make recommendations as to what types of cable should be used with their products. Some coaxial cable is intended for indoor use only, while others may be used *with care* outdoors. By this we mean that any cable, regardless of what it was designed to do, will induce great losses into any system if moisture is allowed to get within its protective outer layer. If connectors are improperly attached to the line or if tears are left unpatched in the plastic outer insulator, the line will quickly become infiltrated with moisture and reception will be poor. Always take as direct a route as possible between the LNA and the receiver, the receiver and modulator, and the modulator and the television receiver with cable runs. It is not a good idea to coil the coaxial cable midway through the line. This adds length to the signal path and causes greater loss.

After years of use, even the best coaxial cable can offer degraded performance. You should plan to replace all cables after a specific period of

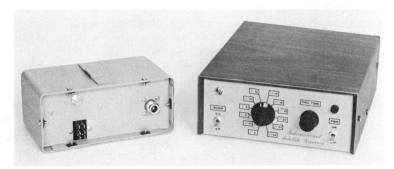

Fig. 3-10. Here, the receiver is shown with a remote control option.

time. Manufacturers' ratings and specifications can help in this area, but a received picture which steadily gets worse and worse over days, weeks and months of operation is a pretty good indication of cable problems.

The connectors (devices which attach the coaxial cable to the various system components) can also present loss problems. Make certain that you use only high quality hardware in this area and that all connector-to-cable attachments are made in accordance with the manufacturer's specifications. Some people have been known to wire connectors to cables in an expert manner and then to attach the connectors to the mating fitting of a piece of equipment poorly. Make sure all connections are tight. A loose connector will allow for moisture to invade the system, bringing along the inherent losses that we are trying to avoid.

One other area of loss which is dealt with in another chapter is that produced by antenna movement. The signals from the satellite are very narrow in comparison with normal broadcast transmissions. We have already discussed obtaining a window on satellites and the importance of a good, properly aligned path between satellite and dish antenna. If your antenna is not mounted properly, even moderate winds can blow it slightly off track. This will degrade reception and possibly eliminate it entirely. The mounting structure must be firm to prevent the antenna from wobbling back and forth, which can cause the picture to fade in and out. There are several critical points in a TVRO Earth station which can affect reception and which do not apply (or at least not as much) to vhf reception of standard television broadcasts. Antenna stability is one of these critical points.

THE ENJOYMENT OF A TVRO EARTH STATION

A person does not get involved in personal satellite reception in order to go through an endless string of headaches. The previous information may have given the indication that microwave reception is fraught with all kinds of perils, which only an engineer could straighten out. This is true, but fortunately, the engineers who designed the receiving equipment which is presently available have worked out most of the difficulties and good basic construction techniques on your part will do the rest. TVRO Earth stations are fun to put together and open a whole new world of entertainment and information on the same television receiver that you have been watching your favorite shows for the past fifteen years. The installation of the antenna and its mount requires the greatest amount of space, but this can be easily provided for in the front or backyard of most homes. Even apartment dwellers may be able to use the roof of their building for such an installation if adequate lot space is not available.

Due to the complexity and critical nature of the circuits used to detect satellite transmission and to make them usable by standard television receivers, the home construction of satellite receivers, LNAs, and other microwave portions of the system is not possible for the average experimenter. A satellite receiver kit is being offered by a manufacturer of completed TVRO products and more on that will be included in a later chapter. This is quite a different situation, in that all of the critical wiring is done at the factory and the builder simply assembles the parts into a working unit. Some persons who have had success in building simple transmitters, receivers, and hobby devices from the many project books available to the electronics experimenter today have often wondered why more microwave kits are not available. The reason for this is that it requires a lot of experience in microwave technology to properly build an LNA or receiver from the ground up. Sure, almost anyone can buy the parts needed to build one, but few people are qualified to perform the delicate and exacting steps required to arrive at a working unit. In order to build microwave cavities used at the feed horn, a knowledge of metalworking is required, along with extremely close tolerances.

For today, the best approach to building your own personal Earth station is to buy the premanufactured antenna, LNA, and receiver. It would certainly be possible to construct the modulator at home, but these devices are relatively inexpensive and can probably be purchased for about the same amount of money that it would cost to build one.

If you're an avid home builder, don't be dismayed. There is still an awful lot of assembly required in putting one of these stations together from the manufactured major components. Often, the dish antenna comes in several pieces and must be assembled at home. One antenna which is discussed in a later chapter requires a great deal of construction technique in order to assemble the kit of parts. But to build the same antenna from discrete components bought at random rather than in the form of a kit with all struts and mounting hardware included and properly cut would be an enormous task. The assembly of these antennas reminds the author of many of the building experiences he had in putting Heathkit equipment together. All of the parts are there, along with the instructions, but it still requires a lot of assembly.

For those readers who detest putting anything together, take heart, because many of the manufacturers offer packages where all components are delivered to your prepared site and set up by technicians. The result is a correctly operating and guaranteed Earth station with no physical labor on your part. The bill for these services tends to be a bit high, however.

QUESTIONS AND ANSWERS

There are a few basic questions which persons ask first about satellite reception and Earth stations in general. Most of these have already been answered by the previous discussion; however, a bit of a recap at this point is in order.

Question: Can I use my present television set to receive signals through the Earth station?

Answer: Yes, you can. The Earth station effectively serves to receive the satellite signals and to convert these microwave frequencies to those your television receiver was originally designed to pick up. If you are going to go to the expense of installing a TVRO Earth station, you will certainly want to have a good television receiver which is in good working condition. The Earth station will in no way improve the basic performance of your present set. In other words, it will not make up for any internal deficiencies in your TV receiver. Persons with older sets which do not offer uhf tuning will want to specify to the manufacturer the inclusion of a modulator which offers an output on vhf (channels 2 through 13). Normally, the modulator will relay the signal on channel 2 or 3, although some units offer the option of choosing any vacant channel desired.

Question: After the Earth station is installed, can I still receive standard television broadcasts?

Answer: Certainly. The satellite reception is accessed by tuning your television receiver to one prearranged vhf or uhf channel. When you change channels within the satellite system, this is accomplished by turning a separate control housed within the satellite receiver or remote receiver control unit. If you desire to receive standard broadcast channels, all that is necessary is to flip the channel control on your television receiver to the station desired. Your set will operate exactly as before, but you will have a special channel which is used for satellite reception only.

Question: Does my present television set require any modifications?

Answer: Usually not. Occasionally, interference from radio sources and power lines may be picked up by the TVRO equipment. When this occurs, the attachment of a small filter at the antenna terminal on the back of the set will remove or reduce this problem. No internal modifications whatsoever should be required.

Question: What kind of programs can I receive from the satellites?

Answer: To give a complete answer to this question would require enough pages to fill up a small book. A general answer would include first-run television programs, uncensored movies (without commercials), sports, stock market reports, all new stations, religious programs, live coverage of congressional meetings, children's programs, and possibly even Russian spy satellites (although the received information will be incomprehensible to the average person not related to James Bond). These are just a few of the programs you can expect with a properly operating TVRO Earth station.

Question: In what areas can I receive U.S. satellite signals?

Answer: United States satellite signals may be received in most of North America, Central America, and a portion of northern South America. In general, all areas of the Continental United States are within a window of at least some of the orbiting satellites. In certain specific locations, local obstructions, such as mountains, steel towers, and tall buildings may interfere with reception. In these instances, moving the installation site to a clearer point a short distance away will completely open the transmission window.

Question: How can I find out which satellites may be received in my area?

Answer: An entire chapter is included elsewhere in this book on this subject alone. Most manufacturers of TVRO Earth station equipment will provide you with a computer printout showing you in which direction each satellite is located with regard to your area. The printout will indicate, based upon your latitude and longitude the number of open windows which exist in your area. These windows are then matched with a list of satellite positions. From the two lists, you can determine which satellites are available to you and which are not. Many computer printouts will also provide an indication of roughly how strong a signal you can expect in your area. This will help in the determination of which receiving system you purchase, especially in regard to antenna size. These printouts are usually provided on a charge basis, with the money often deducted from the price of the equipment should you decide to purchase it from the manufacturer providing the printout.

Question: Do I need a license from the Federal Communications Commission to install my own TVRO Earth station?

Answer: If you intend to operate your Earth station for your own personal non-commercial purposes, you do not need a license from the Federal Communications Commission. The manufacturers of this equipment are required to have their designs type-approved by the FCC before offering them for sale. This assures that no interfering emissions from the receiver or modulator circuits are allowed to cause problems with other television users. Legally, the FCC states that you can own and operate a receiver-only satellite system and you are not required to have a license. However, if more than 50 dwellings are involved and they are not under common ownership or control, it will be necessary for you to register as a cable company. In either case, you will need to obtain permission from and in some cases pay a moderate fee to program owners. For example, The Movie Channel, a popular satellite TV entertainment channel, charges $3.75 per month per user for affiliates or commercial Earth station users operating as their own cable company. However, The Movie Channel does not have an official policy for individual Earth station users, so at present, there is no charge. This means that you can legally receive The Movie Channel at no cost whatsoever. The USA Network, which offers sports and a lot of other interesting programs, charges 11 cents per month for each commercial user but nothing for individual users. Individual users are those of us who own personal TVRO Earth stations. One of the channels which does require the individual to pay a fee is WTBS, Atlanta. The total cost of receiving The Super Station is 10 cents per month per user. ESPN charges 4 cents per month per user. Obviously, the rates are not high, even if you wish to receive a multitude of channels. You are required to pay this fee and to notify the programming company that you are receiving their signal. Sure, it would be easy to cheat and receive them without anyone ever knowing the difference, but it's not fair. Satellite entertainment is something these companies have provided at great expense. In order to continue these services, they obviously must make money. Most users prefer to pay these companies a year in advance, or for that matter, for ten years in

41

advance, if the company will let you do it. Ten years of WTBS would cost a grand total of $12.00 at current rates. Most people feel that the cost is reasonable and helps to assure that we'll always have satellite entertainment available. The appendices in this book contain video program names, addresses, telephone numbers, and contact persons. When your Earth station is installed, a letter or phone call to the program owners of companies you intend to receive will get all of the paperwork squared away.

Questions: What do TVRO Earth stations cost?

Answer: This is a tough one to answer, because it will depend upon your particular area and whether you want a complete installation or one that you assemble yourself. The Downlink System included in this book is one of the least expensive (installed and fully supported systems) presently available and sells for between $5,000 and $7,500. Other installations can cost between $10,000 and $20,000. Various kits are available which include the antenna (which must be assembled by the owner), the completed LNA, receiver and modulator for about $3,100. By shopping carefully and doing some of the assembly work yourself, you should be able to erect a completed system for $4,000 or less. You can easily spend more if you live in a marginal signal area.

Question: What kind of people install TVRO Earth stations?

Answer: The same people who drive automobiles, motorcycles, and airplanes. The same people who own houses, live in apartments, and own thousand-acre estates. The same people who are amateur radio operators, CB enthusiasts, and home experimenters. The same people who are housewives, blue collar workers, and corporate presidents. All types of people are installing their own TVRO Earth stations. The setups are certainly not limited to only the technically inclined, although these persons will probably have an easier job should they elect to do their own installations.

Question: Do I have to use special tuning methods in order to receive satellite signals through my Earth station?

Answer: Not really. Once the Earth station is properly installed and connected to your television set, all standard tuning is done at the television receiver in a normal manner. It will be necessary to rotate and change elevation on the antenna whenever you desire to receive signals from a different satellite. This may be accomplished by electric control with the master panel located in a unit which is placed on top of the television set. Alternately, hand controls are installed at the antenna mount which allow these same adjustments to be accomplished manually.

Question: Don't you have to change satellites each time you want to receive a different program?

Answer: Definitely not. One satellite may transmit twenty or more different channels. The different transmission frequencies or channels are called transponders. On transponder 1, you will pick up a different channel than on transponder 2. Each transponder transmits at a different frequency. To access the various transponders, you simply change channels on the

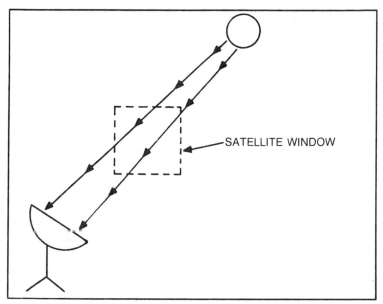

Fig. 3-11. A TVRO Earth station antenna will tune signals through one satellite window at a time.

satellite receiver control. This is exactly the same way you change channels on your home television at present.

Question: With all those satellites up there transmitting on the same frequencies, isn't there a lot of interference?

Answer: No. As has been previously stated, it is necessary to have your antenna in almost perfect alignment with (pointing directly at) the satellite whose signals you wish to receive. When this alignment is slightly off, you get no reception. While there are many satellites in orbit and all transmitting on the same frequencies, your antenna discriminates against unwanted signals by not being directed to an open reception window. Your antenna will tune signals through only one window at a time. This is shown in pictorial form in **Fig. 3-11.**

Question: Don't TVRO Earth stations require a lot of service?

Answer: No more so than does your present television receiver. As is the case in most industries, there are good products and there are bad. I have attempted to present and discuss only the products of reputable manufacturers who are well known in the TVRO Earth station field. If proper care is exercised in making the installation and the preventive maintenance instructions from the manufacturer are followed closely, then your Earth station should provide you with many long years of excellent service. Naturally, if you are located in an area with severe climatic conditions, the wear and tear on the exposed components will be greater than if you were situated in a milder area. Special precautions may need to be taken in those areas which experience unusually high winds, hail, snow, or which are subjected to water-borne salt (sea coast areas). Of course, internal difficulties can develop with Earth station circuits. These will usually dictate the necessity of sending the defective unit back to the factory for repair. Your local television shop won't be able to handle this job for you.

Question: Can I connect more than one television in my home to the TVRO Earth station?

Answer: Definitely. All that is required is the attachment of a standard 70-ohm splitter to the incoming TVRO line. This allows for several televisions to be operated from a single input. The splitters may be purchased at your local Radio Shack or other dealer along with the cables needed. The splitter serves to divide the signal, so remember that you will have half the signal strength at each television when operating two sets from a single line. Splitters which offer outputs for three televisions provide a third of the normal signal strength at each set. If the signal coming from the main line is marginal, you may experience some reception difficulties. This would be true only if there was a very great distance between the modulator and the television sets. This will be encountered very rarely in most TVRO installations. Signal strength to the set is provided by the modulator. The strength of the satellite signals will not be directly affected by the inclusion of splitters.

Question: Can I use my video tape recorder with a TVRO Earth station?

Answer: Yes, you can. The recorder will be hooked up in the usual manner and will function in exactly the same manner as it does with standard television stations.

Question: Don't TVRO Earth stations present a lightening hazard?

Answer: No. As a matter of fact, your present television antenna is more of a hazard regarding lightning hits than would be a typical Earth station. Most Earth station antennas are located near the ground because of their size and weight. Standard television antennas are mounted high in the air. Also, the antennas used for satellite reception are usually mounted on metal platforms whose supports extend into the ground. Lightning hits should not be of undue concern to anyone contemplating the installation of a TVRO Earth station.

Question: Will the Earth station improve my standard television reception?

Answer: No. The two systems are completely different. If you have poor vhf and uhf reception due to your location (and not because of the condition of your television set), it will be just as poor after the Earth station is installed. Your reception of satellite signals, however, will be excellent.

Question: Will the quality of reception using a personal Earth station be as good as that which I obtain from my cable company which has its own Earth station?

Answer: This will depend upon a lot of different situations. If you are located in a marginal signal area, a very large dish antenna might be required to match the picture provided by your cable company. Chances are, they have a very large dish as well, which is typical of commercial installations. A computer printout of satellite windows and signal strengths in your area will help find the answer. For most areas, chances are very good that your personal Earth station, if constructed from good components by a reputable manufacturer, will compare favorably with the cable company. The received image may even be better.

Question: Are there any electrical hazards presented by a TVRO Earth station to small children and pets?

Answer: No more than would be presented by an outdoor lighting fixture. Power to the receiver and LNA is usually derived from the 115-volt house supply. No high voltages are involved in most installations. Owners would do well to restrict access to the Earth station area, as the science fiction look of the antenna is bound to attract the curious of all ages. Unfortunately, the Earth station will probably be a very appealing target of vandals who would take delight in reducing your expensive setup to rubble. For this reason I recommend backyard installations which are generally out of sight of most passersby.

Question: Is it necessary for the antenna to "track" the satellite?

Answer: No. This question is often asked by persons who have seen radar antennas in operation. The satellites we wish to receive are in geostationary orbits. This means that they are always at the same point in the sky. They rotate with the Earth and do not move as far as the antenna is concerned. Actually, the antenna automatically tracks the satellite. As the orbiter moves, the Earth which the antenna is mounted on moves with it. Therefore, you can receive the satellite signals at all times of the day, night, or year without ever moving the antenna in relationship to the Earth.

Question: Can weather affect reception?

Answer: Yes. Heavy moisture content in the air can interfere with satellite reception. During periods of heavy downpours of rain or snow, you will most likely notice a bit of signal fading. Freezing rain can accumulate on the dish and feed horn, interfering with reception and causing structural stresses. Areas which are plagued by these weather conditions often see TVRO installations where the antennas are protected by large non-conducting covers called radomes. This cover in no way interferes with normal reception. Heavy wind conditions can cause the antenna to vibrate. Sometimes, this will cause the signal which is received at the television to fade in or out or to flutter. Most antennas, when properly mounted, will provide normal reception in winds up to fifty miles per hour. They have survival ratings in winds of over 100 miles per hour in most cases.

Question: How long will it take to assemble and make operational a typical Earth station?

Answer: This will depend upon which system you purchase. Preparing the Earth station site can take a week, but most of this time is allotted for allowing the concrete footers for the base to set. The actual work involved in the site preparation can probably be completed in a day and certainly in a weekend in most locations. If you purchase a system which requires antenna assembly, this can take most of a day in some cases and possibly several days when larger or more complex antennas are chosen. Once the antenna is installed on its base, the rest is relatively simple. Interconnecting wiring between the low noise amplifier and the receiver is attached. Another cable is connected between the receiver and modulator (if these are two separate units). A multiconductor cable is run from the receiver to the television set when remote control tuning options are included. This usually takes very little time, especially when there is only a short distance between the antenna and the receiver. As a general rule, it would be fair to state that a full weekend of assembly and adjustment should produce a fully functioning Earth station.

Question: How many people are required to set up an Earth station?

Answer: Again, this will depend upon the system chosen. Generally, it will require three persons to lift a parabolic dish antenna onto its mounting platform and to make the proper attachments. While one man can prepare the site for the antenna mount, two persons can get the job done a lot faster. After the antenna is mounted in place, one person can usually handle the rest of the assembly.

Question: What are the chances of my personal Earth station becoming obsolete?

Answer: Certainly, improvements in TVRO Earth stations will follow the pattern of the entire electronics industry, but manufacturers are working within a range of frequencies which have been established for satellite transmissions. Present receivers are designed to detect signals within this preset range. It is conceivable, although not likely, that the frequency range could be extended, but certainly, modifications to present equipment would be made available to owners. In making a comparison of Earth stations to citizens band radio, we find that when the FCC expanded the CB frequency range to allow for 40 channels of operation instead of 23, the older 23-channel units were not obsolete in that they still provided the same effective communication within the frequency range they were designed for. Your present Earth station would still continue to do exactly what it was designed for even if the satellite frequency range were expanded in a like manner. Right now, these receivers cover the entire entertainment frequency range. If an extension occurred, they would still continue to cover the same range, but you would not have access to the new range without purchasing another receiver or modification kit. Your present TVRO Earth station should provide the same practical reception for many years to come.

Question: What kind of maintenance will need to be performed on my Earth station?

Answer: A properly designed TVRO Earth station requires a very minimal amount of maintenance. All cables should be checked periodically for wear and moisture infiltration. The connectors should be inspected for the same problems. Mounting hardware at the antenna will need to be tightened occasionally and should also be treated with a protective compound in areas near the ocean where salt can be a problem. Of course, you will want to remove any dust which can accumulate within the receiver housing if this unit is mounted out of doors. Weather conditions will play a large role in determining how much maintenance is necessary. After a heavy wind, for example, you might check for loose bolts at the antenna base. As was previously mentioned in this chapter, it's a good idea to remove snow which has accumulated on the dish and to protect the antenna as much as possible during icing conditions. Most maintenance is approached on a common sense basis. A properly installed TVRO Earth station should be easier to maintain than a power lawn mower.

Question: Can my Earth station be protected by my present homeowner's insurance policy?

Answer: Yes, in many instances it can if it is permanently attached to your property. The rates and coverage will vary from area to area and will depend upon such criteria as local weather conditions, crime and vandalism rates, and the amount of coverage you want. I talked with a local insurance agent about this matter and was advised that an Inland Marine policy might be best because of the broad coverage offered. Generally speaking, Inland Marine policies provide insurance on an all-risk basis. Some areas may require certain exclusions. It is difficult to provide a cost figure on this insurance which would be applicable to all areas of the country, but total cost of insuring a $6,000 Earth station on an all-risk basis should come to appreciably less than $200 per year. When contemplating the installation of a TVRO Earth station, a check with your local insurance agent is strongly advised. More than likely, he will want to inspect the installation to make certain that proper construction techniques have been used throughout.

Question: Suppose I move from my present home? Can I take my Earth station with me?

Answer: Certainly. The only thing you will have to leave behind is the concrete footers. The dish antenna is simply removed from the base, along with all electronic equipment. The base is unbolted from the concrete footers and the entire station can be moved in a small trailer. It may be necessary to disassemble the dish if one piece construction is not used on this component. It's a good idea to save the original packing carton in which your equipment arrived. When moving, these will help assure that no equipment damage occurs in transit.

Question: Suppose I move to an area which offers marginal signal reception? What can be done to make my station more sensitive?

Answer: There are many things that can be incorporated in a TVRO Earth station to improve reception in marginal areas. Most of these, however, are rather expensive. The addition of another LNA will bring about some improvement, but the best method is to use a larger antenna. Fortunately, some manufacturers who offer TVRO antennas also offer an extender kit. The Wilson WFD-11 is a good example. This is an 11-foot dish (3.35 meters) which offers a gain of 40.1 dB. If more gain is required, an optional enlargement kit can be ordered which includes extra panels that attach to the present antenna. When the project is completed, you will have a 13-foot dish (4 meters). This offers slightly greater gain which is amplified at the LNA. This is far less expensive than purchasing an entirely new and larger antenna.

Question: Can I mount a TVRO antenna on my roof instead of on the ground?

Answer: This will depend upon the structural strength of the mounting surface. There is absolutely no reason to mount a TVRO antenna high off the ground unless this is necessary to clear large obstructions. It would be wise to consult a local contractor and have him look closely at the intended mounting site and advise you as to what would be required for a safe and durable installation. A typical TVRO antenna and its base can weigh 500 pounds or more. When struck by high winds, the pressure exerted on the assembly can be six to ten pounds per square foot. Your contractor will have to take this into consideration. Again, while rooftop mounting is possible, it should be used only when it is impossible to provide good reception with an Earth station that is mounted on the ground.

Question: Does my property become more valuable when I install a TVRO Earth station?

Answer: Probably. An Earth station is something you don't find in every backyard. From an appraisal value standpoint, your property will probably not increase substantially in worth. But from a resale point of view, a TVRO Earth station could mean the difference between quick sale and no sale at all. Certainly, the price of the Earth station would be added to the asking price of the home and property if you intend to leave it in place when you move. An Earth station facility can be likened to a swimming pool. For those persons who like to swim, the pool will be a good incentive to buy a certain piece of property. Persons who are interested in receiving television signals directly from satellites (and this includes almost everybody) would probably find your house and property a more attractive buy. Looking at this situation from another angle, the Earth station certainly doesn't decrease your property value as it can be moved when you do.

Question: How about financing?

Answer: As always, financing will certainly be available to those individuals with good credit ratings. Some companies may offer their own financing program or you can arrange a loan at your local bank or lending institution. In these days of high interest rates, it's a wise move to shop around for the best loan rate. Some banks may wish to handle these transactions on a collateral basis, using the Earth station to back up the loan. This is very similar to purchasing a new automobile, with the bank technically owning your Earth station until the loan is completely paid off. It will be necessary for you to get a quote from the manufacturer you have chosen and then make all the arrangements in advance of the purchase date. Financial arrangements may differ from bank to bank and certainly from person to person, but in general, the lending institutions I have contacted have no qualms about loaning money to qualified individuals for the purchase of TVRO Earth stations.

48

The previous questions and answers are typical of the concern of persons interested in personal TVRO Earth stations. There are others, however, which will be dealt with in this chapter. Contech Antenna Corporation, 3100 Communications Road, St. Cloud, Florida, publishes an excellent primer on satellite receive-only Earth stations. Other questions and answers were derived from this manual and are reprinted here with their permission.

Question: Why use satellites for television transmission?

Answer: Satellites permit more economical communications (television, telephone, radio, telemetry, etc.) than ground stations or where hard cable connections are required. One satellite can provide 24 transponders with television entertainment to cover all of the United States. Normally, large United States cities have five or six television stations. The reception becomes marginal from these stations for distances in excess of fifty miles from the transmitting station. In telephony, satellites replace literally tons of copper cable that would normally be required to establish telephone communications across distances of a thousand miles or so.

Question: How do the satellites perform their function?

Answer: Satellites are placed in a geostationary Earth orbit. Geostationary implies that the satellites rotate at the same speed as the Earth. This phenomenon is achievable at an altitude of approximately 23,000 miles directly over the Equator. Once a satellite has been stabilized in its orbit, it is used as a transponder. It receives a signal from an Earth station at six gigahertz, converts it to four gigahertz, amplifiers and retransmits a directional signal back to the Earth.

Question: Does that mean that everyone on the Earth can receive these signals?

Answer: No, the satellite from its position in orbit can see approximately 40% of the Earth's surface. In order to provide a sufficiently strong signal to desired locations, the satellite transmit beams are shaped. The U.S. domestic satellites, for example, point their beams at the center of the United States. The signal strength decreases as you go away from the area where the beam is peaked.

Question: Can I receive U.S. satellite signals outside the continental United States?

Answer: Yes, but an engineering evaluation should be conducted to determine the equipment that will be required for the reception of these weaker signals.

Question: Where are the U.S. satellites physically located?

Answer: All of the U.S. DOMSATs are located on the Equator over the Pacific Ocean. The nearest U.S. DOMSAT is WESTAR III located at 87° West longitude, which is approximately 600 miles west of Equator. The farthest U.S. DOMSAT is SATCOM I located at 135° West longitude, which is approximately 3500 miles west of Equator.

Question: How many operational satellites are there?

Answer: Presently, there are eight U.S. satellites and three Canadian satellites that have video transponders.

Question: How many transponders are there?

Answer: The two SATCOMs and three COMSTARs each have 24 transponders. The three WESTARs and three ANIKs each have twelve transponders, which is a total of 192 transponders. Approximately 55 presently carry some type of video programming.

Question: Why do some satellites have 12 transponders while others have 24?

Answer: The ANIKs and WESTARs provide transmission only in horizontal polarization. They have only 12 transponders each. The COMSTARs and SATCOMs alternate vertical and horizontal polarization. Therefore, they can utilize the same total frequency band more efficiently.

Question: Do we expect more satellites to become operational soon?

Answer: Yes. The Canadians expect to put an additional five satellites in orbit through 1982; DOMSAT's schedule is one for 1981, two for 1982, and four in 1983.

Question: Will all of these additional satellites operate in the four gigahertz range?

Answer: No. Some of these will be used in the Ku band—12 gigahertz range.

Question: Will the higher frequency range make present equipment obsolete?

Answer: Not in the near future. The expected satellite life is presently eight years. With the advent of the space shuttle and platform, life may be extended by "preventative maintenance".

Question: How does the antenna contribute to system performance?

Answer: The antenna must capture sufficient signal to overcome the receiving system noise. The weaker the satellite signal in your area, the larger the antenna needed. The signal strength is defined as a power density—watts/square. The larger the antenna, the larger the area or square. Therefore, more power is focused on the antenna feed for reception.

Question: What effect does improper shaping or surface irregularities have on signal received?

Answer: Both shaping and surface rms detract from received signal strength. For example, surface rms of 0.125 inches causes a signal loss of 1.5 dB, which corresponds to 20% power loss of received power. The net effect is that one could have used a physically smaller antenna to perform the same function, or the cause of picture problems may be the antenna.

Question: What if my antenna is spherical rather than parabolic?

Answer: The spherical shaped antennas do not provide the same focusing efficiency as a parabolic. Hence, you will need a larger spherical antenna to do the same job as a parabolic.

Question: Why do you need to perform a structural installation on an antenna?

Answer: There are two basic reasons: (1) Safety—to prevent wind forces from tearing away the antenna from the installation; and (2) The antennas have "pencil beams" which must be pointed directly at the satellite.

Question: What type of antenna mounts are available?

Answer: The two basic antenna mounts that exist are: (1) the elevation/azimuth, which requires independent azimuth and elevation adjustment; and (2) the polar mounts, which permit the "tracking" of the satellite arc through a single motion.

Question: Which one should I use?

Answer: If one does not plan to change satellites often, the elevation/azimuth mounts are less expensive. However, if one desires to change satellites often or to effect changes remotely with a programmed unit, then the polar mount should be chosen.

Question: What options are available for satellite Earth stations?

Answer: Just about every component of the Earth station has many options. It would be best to discuss each one individually.

Modulators. Depending on the manufacture, it can be a simple channel converter or selectable to any of the vhf channels on your television set. The most commonly used modulators have a switch to put your signal on either channel 3 or channel 4 frequencies. Modulators become more expensive when utilized in commercial installations where they are required to drive a few hundred television sets.

Receivers. Two basic types of receivers are currently available. One type is the 36 MHz bandwidth unit designed for Cable TV. The large bandwidth produces good color quality and picture definition. The larger bandwidth accounts for a higher noise figure. This type is differentiated mainly by the "bells and whistles" on the final units. The second type is the one designed for the home market. The units do not meet any industry criteria other than the ones dictated by the marketplace. The home units in general are smaller bandwidth devices that give up picture definition but have better noise figures (better in marginal C/N areas).

The latest innovation in receivers affects the frequency conversion at the antenna. In this manner, you can use both less expensive and longer cables between the antenna and the receiver. The rf frequency being transmitted from the antenna to the receiver will be at the i-f (immediate frequency) rather than at 4,000 megahertz.

LNA. The low noise amplifiers are described by a noise temperature. The most common LNA is the 120° K unit which is basically a 1.5 dB noise figure device. Most LNAs have a gain of 50 dB. This device in general establishes the signal to noise figure of the receiving system. Lower temperature LNAs are used to improve marginal system performances. Before spending money for a 100° LNA, it is suggested that signal strength in your area be determined from a computer printout. If it is indicated from this sheet that a lower noise figure is needed in the low noise amplifier, then the extra expenditures can be made.

Question: Who can help me decide on the options?

Answer: It is suggested that you rely on your dealer for the technical parts of the system, such as LNA temperature, size of antenna, etc. The remote control options are really a function of your personal desires and budget.

Question: Why do you need different size antennas for different parts of the United States?

Answer: As previously mentioned, the satellite antenna is pointed at the center of the United States. The signal decreases as you go away from the peak of the beam. A "footprint" is developed for signal strength.

Question: How do I know what size antenna I need?

Answer: Again, the company you buy your TVRO equipment from will be in the best position to aid you in your selection. Additional information can be garnered from computer printouts for satellite and signal losses in your area. In general, you will want to choose an antenna with the highest gain (largest dish diameter) which you can afford and which is reasonable for your area.

Question: Is an Earth station right for me and my needs?

Answer: You will have to answer this one for yourself. You have probably viewed some of the programs offered from satellites, possibly on your local cable system. Chances are, you have obtained program listings from several different satellite users. You must ask yourself, "Will the pleasure and use I get from my Earth station justify the expense of purchase and installation?" Let's assume your Earth station costs about $4,500. This is about 1½ times the price of a good, large-screen television projection system, and about 6 times the price of a good color television set. Do you have small children in your home who could take advantage of the many educational programs offered? Would your occupation be benefitted by the 24-hour news and information channels available? Do you live in a rural area where first-run movies are shown at local theaters six or seven months after they are released? Do you watch a lot of television? Do you have cable television access or do you live in a fringe area where only a few stations can be marginally received? All of these and many other questions must be answered by the only person qualified to do so—you. I think that a thorough self-examination of your motives will greatly aid you in determining if now is the time to install your own personal TVRO Earth station.

SUMMARY

The expression, "There's a whole new world out there" does not adequately describe all that is to be had with a personal Earth station. The signals are beamed to you from far out in space. This alone opens up many, many possibilities. Experimenters will delight in receiving foreign and military communications satellites and when the experimenting stages are over with, you can sit back and relax with the sights and sounds of a first-run movie.

The installation of a properly functioning ground station is not all that difficult. It can be performed by anyone who has just a bit of mechanical or electronic aptitude. Those persons who have no inclinations in these areas but who desire to receive satellite pictures can contract with Earth station manufacturers for a turn-key operation.

Most owners of Earth stations will be surprised at how fast they adapt to that Martian-looking spectacle in the backyard and to its many uses. At first, the reception of satellite signals on your television receiver will be a sort of anti-climax to all of the planning and work that went into the Earth station installation. But as the user becomes more familiar with all that is available from those transponders far out in space, a universe of possibilities will be opened to you, and all from the comfort of your living room.

It is my firm belief that in one decade satellite television reception will be the rule rather than the exception. Broadcasters are intrigued with this system, which opens up an even larger number of possibilities for them. As development continues in this area, the public should be offered more and more services and satellite use and interest will grow more intense. Satellite television reception will most certainly be the most startling development which has ever occurred in the home entertainment field.

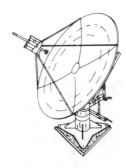

Chapter 4
Obtaining
Surplus Components

It is beyond the scope of nearly every TVRO enthusiast, regardless of technical background, to construct a complete Earth station from the ground up using discrete components to build the many complex circuits. This is discussed elsewhere in this book and alludes to the highly critical nature of frequency-determining circuits at microwave frequencies. Today, when persons build an Earth station, they generally do so by assembling or interconnecting manufactured devices, such as receivers, LNAs, and modulators. Unless you elect to have the manufacturer's technicians come to your site and completely install the Earth station you have purchased from the, you will be doing a major part of the building, which involves antenna site preparation, mounting and interconnecting the devices needed for reception.

A later chapter in this book overviews some of the few kits which are available to the TVRO enthusiast for building TVRO receivers, LNAs, etc. There is another way to go, however. You may be able to obtain some components and completed circuits through surplus channels which can be modified to serve as a part of an operating TVRO Earth station.

Admittedly, the surplus market at present is all but devoid of devices which can be directly applied to satellite Earth stations. One area where this does not especially apply, however, is in regard to dish antennas. You have probably already noticed that the parabolic antennas used for TVRO Earth station purposes closely resemble those which have been used for many years in government radar applications. While there are many different surplus radar antennas to choose from, only those with diameters of ten feet or more are practical for our purposes.

By consulting the catalogs offered by government surplus suppliers, you may become lucky and find one that is suited to modification and will save you many hundreds of dollars over the cost of a commercially manufactured TVRO dish. What we are mainly concerned with when selecting a surplus dish is the diameter of the reflecting surface. The feed horn and any waveguide which may come with it will probably have to be discarded and replaced with components designed specifically for TVRO purposes. These are not excessively priced from most outlets. I have seen feed horns advertised for $25.00 from TVRO companies.

Another important consideration is the mounting base atop which the parabolic dish is situated. You probably won't find too many of these, but it pays to check with suppliers to see if anything might be available. I have used many of the products offered by Fair Radio Sales, which is one of the leaders in electronic government surplus. While their catalogs list many thousands of items, a phone call to this outfit with a specific request for items not offered in their publications is often productive. With the ever increasing interest in satellite television reception, it would be a fair assumption that many surplus outlets will begin concentrating on this growing market.

There are many ancient radar receivers and auxiliary equipment that are offered for sale by government surplus outlets. Most of these are leftovers from World War II, although the Vietnam War has caused newer devices to begin appearing. The older circuits are nearly all of the tube type, but it may be possible to obtain some of the frequency-determining portions of them to serve as a basis for building a receiver for TVRO applications. Certainly, some experimenters will develop conversion designs for a few pieces of military equipment which can be applied to your TVRO earth station.

Another possible source of surplus parabolic dishes will be your local phone company. This source is not often thought of by most experimenters but can offer a wealth of discrete components and entire circuits which can serve many purposes. Did you ever stop to think what happens to all of that sophisticated telephone company equipment when it is replaced by newer models? Chances are, all identifying placards are removed and it's hauled to the local junkyard and tossed on the heap. Some obsolete equipment may be retained for parts salvage, but most is simply discarded.

Being an avid amateur radio enthusiast, I often check with my local phone company to see if they have any "junk" they want hauled away. It is necessary to check with a ranking supervisor in most cases. Some of the "junk" which has been carted to my home has included sophisticated radio teletype machines, high-speed line printers, innumerable printed circuit board cards, and other nameless devices for which there are presently no known uses. The circuit board cards alone often contained parts which would cost well over $100 if purchased separately. Many of these cards were spares that had never been used and were discarded when the equipment they were designed to fit became obsolete.

Telephone companies depend heavily upon microwave communications (including satellite applications) to make worldwide phone service possible. Surely, many of the components used in these applications will become outdated and will have to be discarded. For this reason, it pays to get in touch with an official at your local phone company and let him know of your interests in their obsolete microwave antennas and communications equipment. It may be some time, but you may eventually receive a phone call asking you to haul away some of their "junk" if you're still interested. This can be a two-way benefit. First of all, you can obtain a lot of exotic playthings; and secondly, the telephone company won't have to go to the expense of having their obsolete equipment trucked away.

At present, I have not been able to obtain any dish antennas through telephone company outlets, but have been informed that some may be available in the near future. Since my desires have been made known to the equipment supervisor, there is a good chance that I may get first choice when equipment becomes available. If you're interested in pursuing the surplus conversion route for TVRO Earth station equipment, contacting your local telephone company should be your first step. If you don't know exactly who to get in touch with, the best place to start is the service department. One of their technicians will certainly be able to tell you the name of the person who can give the authorization for you to receive the obsolete equipment. Additionally, (and this is most important) try to establish a good relationship with this technician, because he could be of great help in informing you as to what might be available in the near future. The person who must give permission for you to receive the various pieces of equipment will undoubtedly be quite busy, and your request will not be one of his top priority items. However, when your local technician friend lets you know that a piece of equipment is being junked, you can wait a week or so and call the company official again. This may be enough to jar his memory and get you the equipment you want.

Your local cable company is another source of future surplus equipment. This is discussed elsewhere in this book. Most of the equipment that could become available from these outlets should be directly applicable to home TVRO purposes. Chances are, the telephone company equipment that you receive will be in reasonably good working order, but this will probably not be true of any cable TV surplus which might be available now or in the near future. Cable companies tend to keep their equipment for a long time and have it repaired whenever a defect occurs. After several years of use, equipment repair of defective units may no longer be practical, so new replacements are purchased. Any surplus you would get through them would probably not be operational, but you could probably make it so by doing the work yourself and replacing any defective components. This could be very cost effective, but it will probably be some years before this equipment will be available as surplus in any appreciable numbers nationwide.

Follow the same inquiry procedure with the manager of your local cable company as was recommended in the discussion on telephone companies. It never hurts to ask to be put on a waiting list. The worst they can do it say no.

Another possible source of some limited TVRO components might be had by contacting large corporations which are operative in your area. Near my location such companies as DuPont, General Electric, Western Union, and many others maintain large manufacturing facilities. Some of these use microwave equipment for communications purposes. As old equipment is replaced with new, a possible surplus source exists. Again, it can pay you to make a contact with someone in authority to let them know that you would be interested in removing any surplus they may have. Large companies are often crowded for storage space and may be receptive to such a request. There is a good chance that you would have to pay a small sum for this equipment, but often it's simpler from a bookkeeping standpoint for the manufacturer to simply write it off. This policy will vary from company to company.

In all of these dealings, diplomacy will play a very large role. After all, anyone who grants your request is most likely doing you a favor which could mean a little extra paperwork on their part. I normally make a direct contact by telephone in order to find out the name of someone who might be in a position to help with my request. If possible, the general manager should be contacted. An appointment is then set up at the facility and an "eyeball-to-eyeball" conversation is the result. This is most important, as the person who is in a position to grant your request can learn alot about what you are planning to do with the equipment and may even become interested in the project himself. If you can get his full attention and interest, you have a much better chance of being thought of when something becomes available. This approach is often successful and was borne out some years ago when I contacted a company about some surplus equipment to be used in my amateur radio endeavors. The president of the company became so interested in the project that he paid a visit to my home to see just how his surplus equipment would be used. Not only did I get the equipment, but the president of the company helped me set it up and align it. Afterwards, this same individual expressed his interest in amateur radio and is now an active operator. This will not occur in most cases, but it does demonstrate the possibilities which can be opened by personal contact with individuals within a company which may be able to supply you with surplus equipment. If you are really interested in TVRO Earth stations, it should be relatively easy to get others to feel the same way.

Government surplus was mentioned earlier in this chapter and is available from commercial outlets. However, there is another way to get this type of equipment which involves eliminating the middleman—the supplier. The United States Government regularly holds auctions of surplus equipment which can be purchased by sealed bid. A contact with your local government representative will aid you in obtaining these lists. If you make the highest bid on a piece of equipment, then it's all yours and all that needs to be done is to pick it up and haul it away. Sometimes, it's possible to view their offerings, checking for specifications, condition, etc. Be careful when taking this route, because once your bid is accepted, you are obligated to pay for it and to remove it from government property within a specified period of time. If you don't show up, the Federal Government begins charging storage fees in some instances. Most of the time, however, if you do not pay the balance due, the equipment goes back on the market again. This consumes time and additional taxpayer dollars, so don't bid on anything unless you're sure you really want it. The author points this out because as a lad of sixteen, he received a bid sheet which had a piece of equipment listed that he desired. There were other interesting goodies on there too, so just for the heck of it, he penciled in fifty cents as his bid on each one. The piece of equipment he wanted was obtained by a higher bidder, but his fifty-cent bid on one item was a high bid. The author was informed by the U.S. Department of Agriculture that he had purchased a giant egg incubator (probably weighing several tons), and would he please pick it up by a certain date or be charged for warehouse storage. The intervention of angry (but understanding) parents and the fact that the author was under age saved him in this fiasco.

There are additional ways of at least obtaining information on equipment which may become available. An ad in a large newspaper may produce some results from people who have been collecting surplus components for many years. These have a habit of building up into mammoth proportions and filling needed space. You might even try purchasing an ad in an electronics magazine, but this can be rather costly. Normally, the companies which have surplus equipment are advertising their products in the same magazines.

If you are interested in a particular piece of U.S. government surplus, you can probably get complete information on it by writing the U.S. Government Bookstore in Washington, D.C. for a list of the many publications they offer for sale (and sometimes for free) at reasonable prices. Their military publications are often designed for training technicians and will include a wealth of technical data as well as practical operating instructions.

There are many possible outlets for surplus equipment which may be accessed by the TVRO Earth station enthusiast. Unfortunately, due to the nature and newness of satellite TV reception, most will not have alot to offer at present. By keeping in constant contact with these many potential resources, you can be advised and updated as to equipment which may be offered in the future. It might be a good idea to begin hoarding different types of equipment as it appears in order to one day arrive at enough to build an Earth station or improve a present one. Right now, surplus channels are mainly used to obtain obsolete reflectors from large parabolic dish antennas originally incorporated into military radar systems. Some of these may even be available with motorized controls. These alone may be more than worth the low asking price.

Check around to see what's available in your area. Order all of the surplus catalogs that you can. These are advertised in the popular experimenters magazines sold at most newsstands. Certainly, you would have to wait a long time to collect enough surplus to build a complete TVRO

Earth station, but you may find a device or two which can be used in place of newly purchased and manufactured Earth station components. If you can buy a surplus dish antenna for $100 or so, this can mean a savings of well over $1,000 when it is used to replace a parabolic dish offered by TVRO equipment manufacturers.

By keeping yourself up to date as to the offerings of the surplus market as a whole, you will be in the best position to take advantage of attractive deals as soon as they are made available. In future years, the resourceful TVRO enthusiast has much to look forward to in the surplus market.

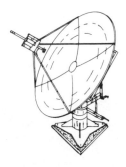

Chapter 5
Principles of
TVRO Antennas

The most prominent feature of a TVRO Earth station is the antenna. This is always a large device, usually at least ten feet in diameter. Most Earth station antennas are generally circular in nature and require sturdy mounts, owing to their relatively high weight. At microwave frequencies, the transmitter radiates power in the form of electromagnetic waves. In satellite television applications, these radio waves are beamed toward a satellite located in a geostationary orbit 22,000 miles above the equator. These transmissions are intercepted by a receiver at the satellite and retransmitted back to Earth again. At the receiving station, the electromagnetic waves must be intercepted and fed to a receiver whose electronic circuits have the ability to remove the video and audio information and pass it along to the television receiver.

An antenna serves to link the receiver with the rest of the universe. Simply put, it is a matching network which couples a transmission line to free space with as much efficiency as possible. Efficiency of an antenna can be thought of as its ability to gather microwave energy and to pass it along to the remaining circuitry. An inefficient antenna system will reflect much of this energy. As efficiency improves, more and more of the energy which strikes the antenna is passed along to the receiver. Of course, there are losses within the receiver system, as well as at the antenna. For TVRO ground station purposes, antenna efficiency would be a comparison of the amount of energy which strikes the reflector and the amount of energy which is delivered to the input of the low noise amplifier.

TVRO antennas are shaped to propagate the electromagnetic waves in a particular direction and more precisely, to a fine focal point. The energy which strikes the reflector is broad in nature or spread out. The dish serves to gather this energy from a large physical area and then combine it in an area which is physically small. This basic principle has been previously discussed.

The most popular type of TVRO Earth station antenna is the parabolic dish. It serves as a reflector at microwave frequencies due to the geometric properties which are obtained from its physical shape. We know that most of the microwave energy which strikes this dish will be reflected to the focal point. This is where the energy actually enters the receiving antenna proper. The reflected wave which is intercepted at the focal point is made up of parallel rays which are all in phase. The antenna beam can be shaped in many different ways. This will depend upon the shape of the dish. Larger dishes produce far narrower beams, while the smaller dishes have broader beam widths.

WAVEGUIDES

The high frequencies employed by satellite television transmissions often necessitate waveguides in certain portions of the receiving system rather than standard dipole or vertical antennas. Waveguides also take the place of standard transmission lines. A waveguide is a metallic pipe which is used to transfer high frequency electromagnetic energy.

Basically, there are two methods of transferring electromagnetic energy. One is by means of current flow through wires. This is the most common method known today, but is far more practical at lower frequencies. The ribbon or coaxial cable which is presently attached between your television receiver and antenna is an example of this type of energy transfer.

The second method of transfer is by movement of electromagnetic fields. The transfer of energy by field motion and by current flow through conductors may not appear at first to be related. However, by considering two-wire lines as elements which guide electromagnetic fields, the current flowing through the conductors may also be considered the result of electromagnetic field motion.

At the microwave frequencies used for satellite reception, a two-wire transmission line is a poor means of transferring electromagnetic energy because it does not confine electromagnetic fields. This results in energy escaping by radiation. Electromagnetic fields must be completely confined. This is where coaxial cable comes into the picture. A coaxial cable consists of a center conductor which is surrounded by an outer conductor. This is shown in Fig. 5-1. The outer conductor serves to contain the energy and greatly reduce or prevent radiation. The two-wire conductor can be thought of as the standard 300-ohm ribbon cable which is often used between a television set and antenna. The coaxial cable is presently more popular than ribbon line, although at standard television frequencies, the ribbon line may be more efficient. Even the coaxial cable is not very efficient at the microwave frequencies we will be dealing with in TVRO Earth stations. Long cable runs can quickly render the entire system useless.

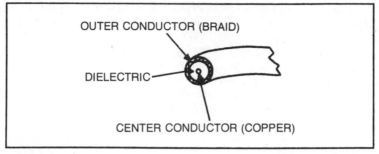

Fig. 5-1. Coaxial cable is made up of a center conductor and an outer braid.

Energy in the form of electromagnetic fields may be transferred very efficiently through a line that does not have a center conductor. A waveguide can be thought of as a hollow transmission line without a center conductor. The configuration of the energy field in a waveguide is different from that in a coaxial cable due to this missing conductor. Waveguides may be rectangular, circular, or elliptical in cross-section.

There are three types of losses in radio frequency transmission lines. These are copper losses, dielectric losses, and radiation losses. The copper loss is due to resistance within the copper conductor. This is often referred to as I^2R losses. I^2R is part of *Ohm's law* for computing power. Power losses are equal to the resistance of a conductor (R) times the current flow (I^2). The conducting area of a transmission line will determine the amount of copper loss. The larger the cross-sectional diameter, the lower the losses. Dielectric losses are due to the heating of the insulation between the conductors of a transmission line. This can occur when a transmission line which is simply too small is used to transmit high amounts of power or when the dielectric does not provide sufficient insulation between the center conductor and the outer one. A dielectric which is a good insulator at 30 megahertz may act like a conductor at 3,000 megahertz (3 gigahertz). Radiation losses are due to the radiation of energy from the line. Coaxial cable usually prevents most of this radiation.

With these facts in mind, the advantages of waveguides over a two-wire and coaxial transmission line can be better understood. A rectangular waveguide with a large surface area is shown in Fig. 5-2. A two-wire line consists of a pair of conductors with relatively small surface areas. The total surface area of the outer conductor of the coaxial cable is large, but the inner conductor is small in comparison. At microwave frequencies, the conductor becomes electrically smaller. This is due to a phenomenon known as skin effect. Put simply, skin effect is a tendency for high frequency energy to flow only on the outside surfaces of the conductor rather than through the entire cross-section. So, coaxial cable used in TVRO Earth stations will restrict a certain amount of energy flow. The size of the energy field is limited by current flow and is thus restricted to the current-carrying area of the conductor.

With this in mind, it can be seen that a waveguide will have the least copper loss of the three types of transmission lines under discussion, because it has no center conductor. Dielectric losses are very low too. This is due to the fact that no solid supports are needed for a center conductor. The dielectric in a waveguide is air. This has a very low dielectric loss at microwave frequencies compared with solid conductors. Radiation losses are negligible in a waveguide as well, since the electromagnetic energy is wholly contained within the structure.

With all of these many advantages, one might ask why waveguides are not used throughout the entire TVRO Earth station facility. We know at this point that coaxial cable is incorporated between the LNA output and the receiver. While a waveguide is used at the LNA/feed horn connection in TVRO Earth station, its physical size and structural requirements make it expensive and difficult to work with. Therefore, we accept the losses presented by coaxial cable in order to save some money and to have a system which is easier to set up and adjust.

The dimensions of a waveguide are critical. They must be approximately a half-wavelength at the frequency of operation. At four gigahertz, the waveguide width would be about 1½ inches. If these dimensions are not adhered to, energy at four gigahertz, for example, and all frequencies below it would not travel down the guide whose opposite end would be connected at the receiver. There is also an upper limit to the frequency which may be transported by a waveguide. Therefore, the frequency range of any system utilizing a waveguide is limited. This is not the case in standard transmission lines, such as coaxial cable. Coax will transport electromagnetic energy over a tremendously wide range of frequencies. But as the frequency increases, losses become more and more abundant.

The travel of energy down a waveguide is similar but not exactly identical to that of the propagation of electromagnetic waves in free space.

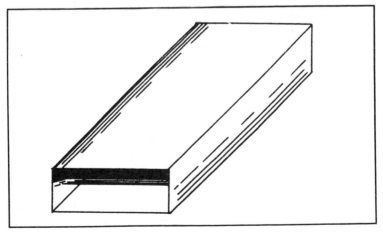

Fig. 5-2. A rectangular waveguide has the advantage of a larger surface area.

The difference is found in the fact that energy in a waveguide is confined to the physical limits of the guide itself.

HORN ANTENNAS

A horn antenna is shown in Fig. 5-3 and provides a method of matching the input impedance of a waveguide to free space. Normally, the receiving pattern of an open-ended waveguide is broad in both the vertical and horizontal planes. The evolution of the horn antenna came about as a result of the effort to minimize the reflections that occur when a straight piece of waveguide receives microwave energy. In other words, a length of waveguide connected to a receiver could serve as an antenna, but it would be a very inefficient antenna because of its mismatch to free space. The horn antenna is actually a piece of waveguide which has been flared at the open end. Figure 5-4 illustrates this.

When a waveguide is flared out, a horn antenna is obtained; but the flare must be very gradual so as to permit a better match between antenna and free space. The horn is very practical at microwave frequencies, since its physical size is not prohibitive. Since a resonant element is not used, this type of antenna is capable of wideband operation.

Another type of horn antenna is shown in Fig. 5-5. This is called a conical horn. It is sometimes seen in TVRO Earth station uses. The previous type is called a pyramidal horn and has equal directivity in both the vertical and horizontal plane.

The reader may be wondering at this point why the discussion has become so dry and technical. The reason is that it is necessary to understand a few basic principles in order to perceive the operation of Earth station antenna systems. The horn antenna is normally used at the focal point of the dish. It is connected to a waveguide at the input of the low noise amplifier. This is a critical point in the antenna circuit, because no amplification of the received energy has taken place yet. We use a very short

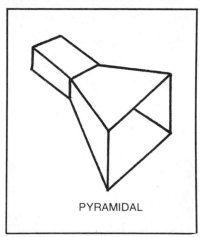

Fig. 5-3. A horn antenna serves to match the input impedance of a waveguide to free space.

PYRAMIDAL

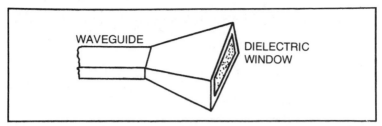

Fig. 5-4. In actuality, a horn antenna is really a piece of waveguide which has been flared at the open end.

length of waveguide between the horn and the LNA, because the use of coaxial cable or some other type of solid transmission line at this point would sharply reduce the amount of signal delivered to the amplifier circuitry. The feed horn/waveguide input to the LNA is extremely efficient and very little of the energy at the focal point is reflected away.

The magnitude of the signal input to the input of the amplifier is increased within the LNA circuitry. Now, we have much more signal to work with, having gotten through the free space portions of the system. Coaxial cable, in short lengths, may now be used to transfer the higher magnitude signal to the receiver circuitry without incurring losses so great as to input a deficient signal to the receiver. Sure, it would be far better to use waveguide between the LNA output and the receiver input, but due to the short distances involved, the price and complexities are simply not worth it.

PARABOLOID REFLECTORS

The paraboloid or parabolic reflector serves as a collecting and focusing device and is an integral part of an Earth station antenna. Since the feed horn dimensions are small, the reflector is large in terms of wavelength. A reflector surface is chosen which will provide a constant phase. This means that all of the gathered energy striking the dish will travel the same distance. For our purposes, this also means that all of the transmissions at any one instant which were sent by the satellite will strike the feed horn at the same time.

Fig. 5-5. A horn antenna serves to match the input impedance of a waveguide to free space.

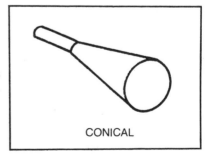

CONICAL

But how can this be? The feed horn is obviously closer to the center of the dish when using a parabolic reflector than it is to the outer edges. It would seem that the signals which strike the outer edges would have to travel farther than those which strike at dead center. The answer is that they do. Signals which strike at the center of the dish have a shorter distance to traverse to the feed horn than those which strike at the edges. This is shown in Fig. 5-6. For the sake of discussion, let's assume that the distance from the feed horn to the center of the dish is 8 feet. Also assume that the distance from either edge is 10 feet. There is no doubt about it; ten feet is a longer distance than is eight feet. But do the signals really travel farther?

Now that you're completely confused, let's attempt to clear this matter up by using Fig. 5-7. To be in phase, the signals must travel the same distance. But this does not mean the same distance from the dish which collects the transmissions to the feed horn, but from the satellite to the feed horn.

Figure 5-7 shows the satellite, the parabolic dish, and the feed horn. Again, the feed horn is located at a distance of eight feet from the center of the reflector and ten feet from either edge. Let's assume that the satellite is located exactly 40,000 miles from the back of the feed horn. The incoming signal from the orbiter must travel 40,000 miles to get to this point. But it must travel an additional eight feet to get to the center of the dish and then eight feet more after being reflected to enter the aperture of the feed horn. Now, we must remember that the parabola is a curve. Therefore, the center of the dish which lies directly in a line with the satellite and the feed horn is the farthest point from the orbiting transmitter in the whole antenna system. The outer edges of the dish arc farther from the feed horn but

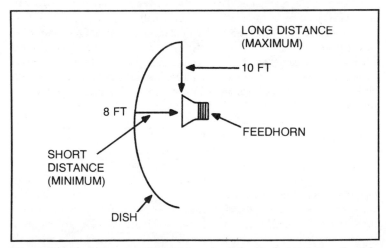

Fig. 5-6. Signals from the satellite travel varying distances from points on the dish to the feed horn.

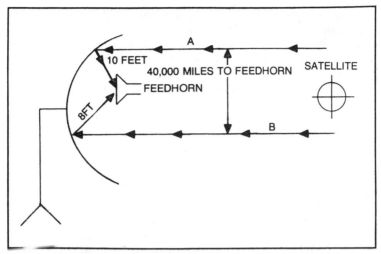

Fig. 5-7. A simplified diagram of the paths which signals must travel from the satellite to the feed horn.

closer to the satellite. By studying the parabola in great detail, we would learn if the center of the dish is 40,000 miles plus eight feet from the satellite, then (in this example) the outer edges must be 40,000 miles plus six feet from the satellite. Let's repeat this again for clarification purposes. To get to the center of the dish, the satellite signal must travel 40,000 miles and eight feet. The same transmission will travel only 40,000 miles and six feet to get to the outer edge of the dish. Forget the feed horn for the moment. The outer dish edges are physically closer to the satellite than is the dish center. The transmission from the satellite, then, does not strike the different areas of the dish at the same time.

Refer to Fig. 5-7 again. Transmission ray A from the satellite strikes the outer edge of the dish at a distance of 40,000 miles and six feet. This ray is reflected toward the feed horn, where it travels an additional ten feet to reach the aperture. Total distance traveled is 40,000 miles and sixteen feet. Now, let's look at transmission ray B. It travels 40,000 miles and eight feet to strike the center of the dish. It is then reflected toward the aperture, where it travels another eight feet. Total distance traveled is 40,000 miles and sixteen feet. This is the exact distance which was traveled by ray A. Again, there is a distance of eight feet between the dish and aperture at the center and a distance of ten feet between aperture and outer edge.

It can be seen from this discussion that while ray A is being reflected by the outer dish edge, ray B is still traveling toward the center. When this ray is finally reflected, ray A has already traveled two feet of the distance to the aperture. It is eight feet away from its target. At this same instant, ray B is also eight feet away from the aperture.

This discussion has used two-dimensional drawings to depict the parabolic reflector, which is a three-dimensional device. If you can conceive

67

of this principle being applied to millions of rays at millions of points on the dish, then you understand the operation of a parabolic antenna.

The operation of a parabolic antenna helps to explain why alignment of the antenna with the satellite is so critical. If the dish is only slightly off the beam, so to speak, the transmitted rays would strike the dish in such a manner that the signals reflected to the aperture would be out of phase. This is shown in Fig. 5-8. Here, the leading edge of the antenna would be the first area struck. The trailing edge would be the last to receive the signal. The original transmission would be reflected to the feed horn over a time span which does not coincide with the time required for the original transmission. For proper reception, a one-second burst of transmitted energy and the signal rays contained therein must strike the receiving antenna (the feed horn) over a span of exactly one second. If this latter period is 1.5 seconds, most of the transmitted information is lost due to the out of phase condition. For this reason, a TVRO Earth station antenna must be sighted with the same precesion marksmen use in shooting at small targets except that your target is thousands upon thousands of miles away.

SPHERICAL ANTENNAS

Another type of antenna which is becoming popular for TVRO ground station applications is called the spherical antenna. Like the parabolic antenna, this design focuses the microwave energy reflected from its

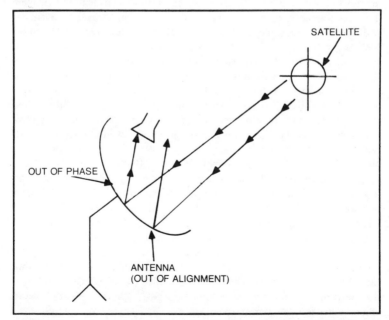

Fig. 5-8. If the antenna is not properly aligned with the satellite, signals will be out of phase.

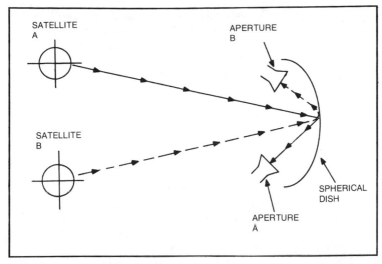

Fig. 5-9. In a spherical antenna, signals from two satellites may be received without moving the antenna.

surface to a point in front of the antenna. The parabolic antenna has one focal point which lies directly in front of and perpendicular to the center of its reflector. For this reason, it must be directly aimed at the satellite which is to be received. This method is often referred to as boresighting. This term was probably garnered from the fact that the parabolic dish is aimed at the satellite in the same manner that a gun is aimed at a target.

There are major differences, however, in the operation of a spherical antenna. This design focuses microwave signals received by as much as plus or minus 20° off boresight (perpendicular to the center) with a negligible loss of efficiency. This means that a spherical antenna does not have to be exactly aligned with satellite, but more importantly, a spherical antenna positioned to reflect and focus the signal from one satellite will simultaneously reflect the signals of others. These must be adjacent satellites within plus or minus 20° of boresight. The spherical antenna reflects these adjacent satellite signals most efficiently but at different points in front of the antenna. In other words, when the spherical dish is boresighted on one satellite, it efficiently reflects these signals to a specific focal point. At the same time, the signals from an adjacent satellite are also being reflected, but to a different focal point. Both points still lie in front of the antenna.

Figure 5-9 shows a graphic representation of the spherical antenna using signals from satellite A and satellite B. It can be seen that the signals from the first satellite are reflected to aperture A which lies more to one side of the reflector than does aperture B. Notice that this is accomplished without moving the reflector. A feed horn placed at aperture A would receive the transmission of satellite A. If you took this same feed horn and moved it to focal point B, satellite B would be received.

A low noise amplifier and feed horn combination properly positioned at either of these focal points will amplify the maximum amount of signal that this antenna is capable of reflecting from a given satellite. This occurs provided that the signal is received at the reflector at less than a 20° angle from the boresight. By moving the LNA/feed horn from one focal point to the other, several satellite signals are received with about the same efficiency and no repositioning of the large dish is required. It should be noted that while the reflector's efficiency is generally constant within the 20° receiving arc, highest efficiency is approached at boresight and to angles either side of this point.

Figure 5-10 shows the Skyview I antenna, which is a 12-foot spherical dish that is put together from a kit offered by Downlink, Inc. This antenna is discussed elsewhere in this book as to construction procedures and performance. We will use this design as an example for the continuing discussion. Since the spherical antenna can reflect and focus the signals from more than one satellite simultaneously and in an efficient manner, two variations of antenna azimuth positioning may be considered.

If you desire to receive only one satellite, the antenna will be set up in the normal manner with its center being directed toward the azimuth heading for that satellite. Once this is done, the LNA and feed horn will be positioned at the focal point which lies directly in front of the antenna's center.

For multiple satellite reception, the antenna will be directed toward the azimuth heading of the center of the group of satellites whose signals can be properly focused by the reflector. This is shown in Fig. 5-11. The LNA/feed horn is then positioned at any focal point for reception of a given satellite in the group. When reception of another satellite is desired, the

Fig. 5-10. The Skyview I antenna is of spherical design (courtesy Downlink, Inc.).

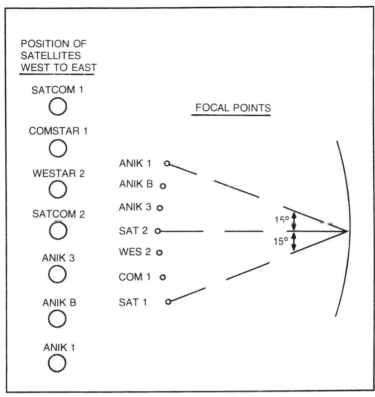

Fig. 5-11. To receive more than one satellite, the antenna is directed toward a centered azimuth heading of the group to be received.

feed horn is moved to another focal point. This, of course, is done manually. The feed horn/LNA is not physically attached to the spheroid reflector but is mounted separately some distance away. The Downlink antenna uses a concrete base to support the feed horn. The base is heavy enough to hold its assembly in place but can be moved by hand to different focal points.

Alternately, two or more LNA/feed horn assemblies may be positioned for reception of multiple satellites simultaneously. A remote switching unit can be used to change from one feed horn to another. Technically, it would probably be possible to incorporate as many feed horns and LNAs as there were satellites within the 20° receiving angle. At $1,000 or more per unit, however, this could get quite costly.

In order to properly position the feed horn and LNA to receive the maximum amount of signal, three factors must be taken into consideration. First is the focal length of the antenna; second, the angular difference between the inclination of the antenna and the satellite elevation; and finally, the angular difference between the azimuth heading of the antenna and the azimuth heading of the satellite.

The focal length of the Skyview I antenna (12-foot model) is fifteen feet. This is the distance measured from the center of the reflector to the center of the feed horn. Since this is a kit design, this distance may vary slightly, depending upon the consistency of the reflector's curvature. Knowing this, we also know that the feed horn will always be fifteen feet from the center of the antenna, and the mouth of the horn will always aim directly at the antenna center, no matter which focal point is used. This is shown in Fig. 5-12 as an arc along which the feed horn may be positioned.

The height of the LNA/feed horn assembly will vary, however. This will depend upon the elevation of the satellites and the antenna inclination. A signal coming into the reflector at an angle which is measured from a line drawn perpendicular from the center will be reflected at an equal angle in the opposite direction. When the feed horn is repositioned left or right in front of the reflector to receive another satellite, its height above the ground must be changed. Figure 5-13 shows how the focal point varies in height for each satellite elevation above the horizon. This height can approach impractical limits with some inclinations, so this must be worked out beforehand.

Positioning of the LNA/feed horn to a focal point in the azimuth direction requires approximately one foot of movement (left or right) for every 4° of change in azimuth. The azimuth focal point of a specific satellite in relationship to the center of the antenna is found by using the formula: feed horn movement from center (in feet) = the satellite azimuth − the

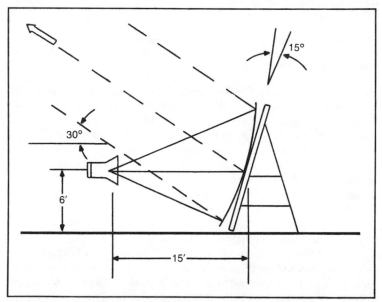

Fig. 5-12. The arc along which the feed horn may be positioned to receive the maximum amount of signal.

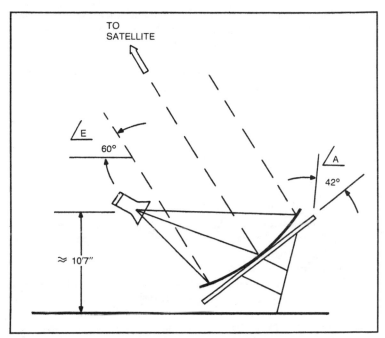

Fig. 5-13. Here, it can be seen that the focal point varies in height for each satellite elevation above the horizon.

antenna azimuth, with the difference divided by four. The LNA/feed horn height is far more complicated mathematically and reads as follows: LNA/feed horn height = 6 cos (angle A) + 15 sin (2 angle A − angle E), where angle A is the antenna inclination and angle E is the satellite elevation. For most of us, working this formula is a little scary on the first try, but if you have an electronic calculator which offers sin and cos functions, the answer can be figured in less than a minute. Fortunately, the manufacturer of the Skyview I antenna provides a lot of calculated information on most satellites of interest to American viewers. Since the formula for left or right positioning from center is simple to work, you could probably set up the antenna for proper receiving on a trial and error basis.

The LNA/feed horn adjustment that will be described was supplied by Downlink, Inc. specifically for their 12-feet spherical dish, but these instructions will generally apply to most spherical dishes if the focal length is adjusted to match that of the antenna being tuned. First of all, the LNA/feed horn is positioned directly in front center of the antenna at a distance of fifteen feet (or at a distance which matches the focal point). Next, the azimuth position is adjusted. This is calculated by moving the LNA/feed horn left or right of center by the appropriate distance figured from the formula. Remember to maintain the focal length (15 feet in this case) at all times. Calculate the required height of the feed horn and aim it toward the center of the antenna. The feed horn will now be within a few inches of the exact focal point. This will depend upon the accuracy of the antenna's curvature and the inclination/azimuth setting.

Now, while watching a monitor, move the feed horn slightly and systematically until a picture is obtained. You will want to make certain that the LNA/feed horn is set up for the proper polarity (vertical or horizontal) for the satellite you are trying to receive. Once a picture has been obtained, it's an easy job to make minute positional adjustments for maximum signal strength.

PARABOLIC/SPHERICAL COMPARISON

Now that the two main types of antennas designed for TVRO Earth station applications have been discussed, it is appropriate to compare the two. This is a difficult task in some ways because they each have equal advantages and disadvantages. In the end, the specific operating needs of the Earth station owner will most likely be the determining factor in deciding which design will be best for his particular needs.

The spherical antenna has an advantage in simplicity of mounting hardware. The antenna is set and then forgotten. Only the feed horn/LNA assembly is moved for tuning. The parabolic dish antennas must be aimed each time a different satellite is to be received. This aiming requires the movement of the entire antenna structure instead of just the feed horn. But from a positive standpoint, the parabolic dish can be controlled from the television receiver location when optional equipment is purchased. Most manufacturers offer motorized drives which are fitted to the parabolic dish base mount. These motors are activated by a control box which is mounted at the television receiver. This certainly adds to the convenience of TVRO Earth station operation. The spherical antenna, on the other hand, requires the physical moving of the LNA/feed horn assembly each time a different satellite is to be accessed. The dish is not physically attached to the feed horn, so rotation is neither possible nor desirable. While two LNA/feed horn assemblies may be used with a single spherical reflector, this becomes prohibitive from a cost standpoint.

In all fairness, it is proper to mention at this point that most TVRO Earth station owners will not be switching from satellite to satellite all that frequently, regardless of what type of antenna system they have. Sure, in the first few weeks of operating the Earth station, every satellite with an open window will certainly be accessed. This would be part of a normal breaking-in period for the Earth station as well as for the excited owner. But after the novelty wears off, the Earth station's operation becomes more in line with what it was originally purchased for—the reception of entertainment channels in a convenient and readily available manner.

Most satellites which carry entertainment programming have many other channels as well. For example, the video services on Satcom I include the following: Nickelodeon, PTL, WGN, The Movie Channel, WTBS, ESPN, CBN, USA Network, Showtime West, Showtime East, CNN, WOR,

Reuter/Galavision, Cinemax East, HTN, Home Box Office West, Cinemax West, Home Box Office East, and several spare channels. That's at least nineteen major programming services. Everything from children's educational programs to religion to first-run movies is included on this one satellite. So assuming you don't like the movie that's playing on The Movie Channel, which is on transponder 5, all you do is switch to transponder 24, where Home Box Office is airing something you consider to be decent, or vice versa. If you get tired of first-run movies, you can always go to CNN, the all news network. Satcom I and other such satellites individually carry a wide enough range of programming to suit almost anyone. The result of all this is to make it unnecessary to go to more than one satellite. This is called competition between programming services.

So, don't place more importance on picking up every satellite in the sky instead of just one or two than this practice merits. Of course, it is very convenient to be able to alter the position of the antenna in order to change satellites. Most TVRO Earth station owners keep track of all the services that are being offered through monthly publications. This gives them the opportunity to check out each program almost as soon as it becomes available. Many TVRO Earth station owners like to experiment with their installations. There are many, many satellites of all types and from all nations in orbits which can be intercepted in the United States. The appendices of this book contain information of many of them. By owning an antenna which can be rotated, many of these satellites can be accessed, although the information they transmit may not be intelligible (this would especially apply to the Russian spy satellites). Persons who try to access any satellite they have a window to can be directly compared with shortwave listeners who are not so much looking for information as they are seeking any kind of transmission which can definitely be attributed to that particular orbiter. They may even keep lists on which satellites they have been able to access and those that are still on the want list. QSL cards which are verifiers sent out by amateur radio operators and other transmitting stations are never seen in TVRO service, but keep trying. All of these activities would not be possible without an antenna system which can be directed to many different satellites.

The choice of a spherical or parabolic dish will most likely depend on the Earth station owner's personal desires. The spherical antenna described in this chapter would receive along an arc plus or minus 20° from boresight. By obtaining a computer printout of satellite windows in your area, you can quickly determine where the satellites you are interested in will be found. If all or most of these could be hit with a plus or minus 20° reception area, then the spherical dish might be a good choice. On the other hand, if you want constant reception over a larger portion of the sky and don't like the idea of going outside to change satellites by readjusting the feed horn, then the parabolic dish would be the way to go. Most parabolic dishes, however, are not fitted with motorized controls. At present, these are the exception rather than the rule in most TVRO Earth stations. So, even with the parabolic dish, you will probably have to go outside to change satellites, although the readjustment may be simpler than that which is necessary for spherical designs.

Another comparison which is necessary is that of cost. This is another difficult parameter to measure, because there are great fluctuations within the market. The spherical antenna is usually offered as a kit of parts and for this reason, it may be cheaper than most parabolic dishes of the same diameter. Parabolic antennas are also offered as kits, but there are just a few parts involved in assembling the latter, whereas the former is a major project. The struts and base assembly for the parabolic dish will be more costly than those for the spherical design, which will most likely be simply placed with the lower edge on the ground and a few simple supports from behind. This again tends to make most spherical antenna designs less expensive than comparable parabolics when considered as a whole system (dish, feed horn, mounts).

There is an aesthetic consideration as well. In my opinion the parabolic dish is neater in appearance. Most structures are rather striking and massive, but the seemingly one-piece design of parabolics provides less of a chicken-wire effect than does the spherical. The spherical antenna usually consists of a fine mesh screen mounted on a support lattice which is visible from front and back. Fiberglass construction is pretty much universal with TVRO Earth station dishes designed for personal applications when the parabolic design is considered. Fiberglass is not a conductor of electricity, but serves only as a smooth-surfaced container for the screen which is sandwiched between two molded sections of the fiberglass.

Another problem with spherical antennas is positioning of the feed horn, especially when the dish is tilted to intercept satellites which lie many degrees from the horizon. When the upper edge of the dish is tilted back, the LNA must be elevated higher from the ground. This can present structural problems, in that it is necessary to provide a mast support for the feed horn which may have to be fitted with guy wires for stability. This tends to add a lot of permanence to the feed horn installation, and changing satellites is much more difficult, in that the guy stakes which anchor the wires to the earth must be removed and the entire assembly set up again at a different focal point.

It can be seen that there are many criteria to be considered when deciding between a spherical and parabolic antenna design. Most companies that sell spherical antennas also offer the more standard parabolic types, and once you explain your location and intended used of the completed Earth station, they should be in an excellent position to give you further advice on your selection. The author has found in talking to many representatives of companies which offer equipment for personal TVRO Earth stations that they are more than willing to provide information and assistance, even if you're not sure that an Earth station setup is for you. They must feel (and rightly so) that the more they disseminate information

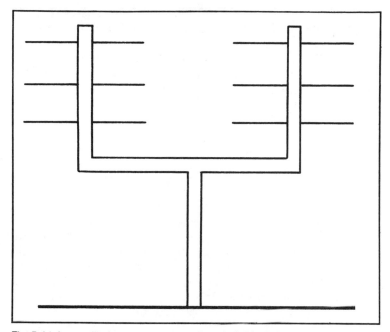

Fig. 5-14. In a multi-element antenna system, individual elements are combined to provide a single output.

about Earth station operation, the more interest will be generated about the field as a whole. This helps the entire industry.

OTHER TYPES OF ANTENNAS

While the parabolic and spherical antenna designs are used almost universally for TVRO Earth stations, there are other types of antennas which do have microwave applications. A multi-element array can be used, but its design is very complex when compared to the relative simplicity of microwave dishes. Figure 5-14 shows a multi-element array which consists of many elements or individual antennas which are combined to provide a single output. The length of each horizontal element section would be a quarter wave at the operating frequency. Two of these sections are combined in the same plane to produce a half-wave dipole. Each element is fed at the center. Though often massive in size, the use of thin wall aluminum tubing makes the overall design very light in weight.

A major problem is encountered with this type of design, in that when the elements are in the horizontal plane, the entire system responds efficiently to only horizontally polarized transmissions. To become vertically polarized, the system would have to be canted 90° to place each element in a vertical position in respect to the Earth. With a structure this

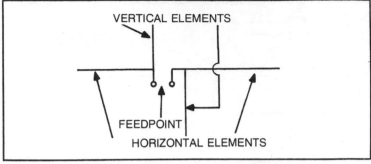

Fig. 5-15. In this design, vertical and horizontal elements are fed from the same line.

large, this type of rotation is not practical. Some designs may incorporate both vertical and horizontal elements fed from the same line. This is shown in Fig. 5-15. Now, the antenna system is both vertically and horizontally polarized and need not be canted on end for vertical reception. This, of course, doubles the complexity of the design and makes the structure about twice as heavy. Wind storms tend to play havoc with these types of designs.

Another type of antenna which has seen some limited usage at microwave frequencies is the quad array shown in Fig. 5-16. This antenna consists of a large number of square elements whose parameters equal a full wavelength each. A single quad antenna element is shown in Fig. 5-17. The feed point is at the bottom horizontal portion of the element. A quad is a mixture of both vertical and horizontal components, so it will respond reasonably well to both horizontal and vertical polarized transmissions. In the quad design, only one element is attached to the feed line. The others

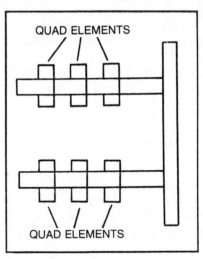

Fig. 5-16. Quad arrays have limited usage at microwave frequencies.

78

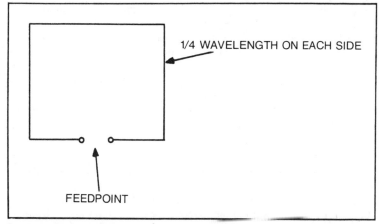

1/4 WAVELENGTH ON EACH SIDE

FEEDPOINT

Fig. 5-17. Shown here is a single quad antenna element, with the feed point at the bottom horizontal portion.

are called parasitic elements and each is cut to a different size. Figure 5-18 shows one section of a quad array. The driven element or active element is the one to which the feed line is connected. The elements to the back of the active one are called reflectors. The perimeter of the first reflector (the one immediately behind the driven element) is slightly larger than the one to which the feed line is attached. The second reflector has a perimeter which is larger than the first. The reflectors increase in size as their numbers increase.

The elements in front of the driven element are called directors. The first director is slightly smaller in perimeter dimensions than the driven element. Each successive director on out to the front end of the antenna is

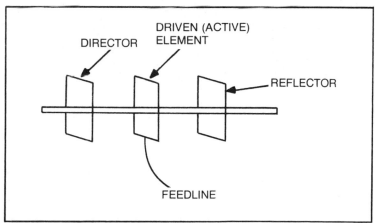

DRIVEN (ACTIVE) ELEMENT

DIRECTOR

REFLECTOR

FEEDLINE

Fig. 5-18. In a quad design, only one element is attached to the feed line, with the others called parasitic elements.

smaller in perimeter size. Several of these sections are combined to form a quad array.

The design of each of the many elements in such an array is extremely critical. At four gigahertz, a full wave element would be a fraction less than three inches around its perimeter. Each reflector would be a small fraction of an inch larger, while each director would be an equivalent fraction smaller. A quad array with substantial gain could easily have fifty or more elements. Any deviation from proper design length at any of these will result in degraded performance. Obviously, attention to detail with this and any other multi-element microwave antenna is crucial.

A similar type of antenna array may be formed from several horizontally or vertically polarized yagi antennas which are combined to form an antenna system. Such an array is shown in Fig. 5-19. The yagi antenna is a design which is most often used in the vhf and uhf frequency spectrums for standard television reception and to receive FM stereo stations. The same principles apply to the yagi as were discussed in the section of this chapter dealing with quad antennas. Each yagi element is about a half wavelength at the microwave frequency. As before, each major section is composed of a driven element, reflectors, and directors. A closeup of one of these sections is shown in Fig. 5-20. Here, the driven element is fed at its center. It is exactly one-half wavelength at the operating frequency. Each reflector is

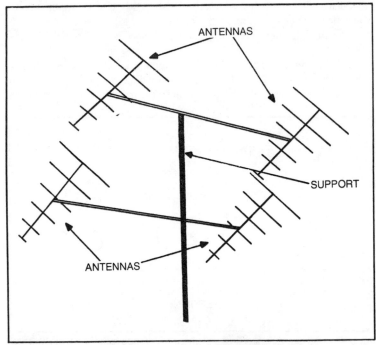

Fig. 5-19. Here, several yagi antennas are combined to form an array.

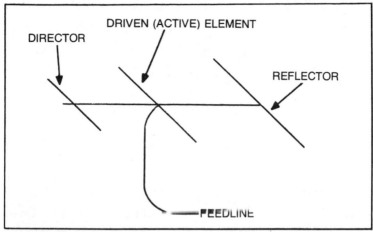

Fig. 5-20. In this closeup view of one of the sections of a yagi array, the driven element is fed at its center.

slightly longer, while each director is decreasingly smaller. As shown, the antenna is horizontally polarized. The entire system would have to be canted 90° to place the elements in a vertical plane in order to be used in the reception of vertically polarized waves.

Still another type of antenna which has seen some usage at microwave frequencies is the helix shown in Fig. 5-21. The helix consists of circular turns of copper conductor mounted in front of a reflector. Each turn is one wavelength long and the spacing between turns is approximately 0.25 wavelength. The turns are supported by a dielectric which gives good ratings at microwave frequencies. The reflecting screen is one wavelength square (about 3 inches by 3 inches or four gigahertz).

Several helix antennas may be combined in parallel to form a helix array. This is shown in Fig. 5-22. The reflector would be four wavelengths square if four antennas were used, six wavelengths square when using six

Fig. 5-21. The helix antenna is made up of circular turns of copper conductor mounted in front of a reflector.

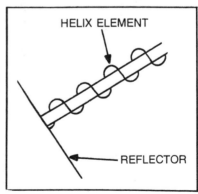

antennas, etc. The total number of turns used to make each of the major antenna elements will determine the overall gain of this design.

The helix antenna is neither horizontally nor vertically polarized. At the same time, it will respond with fairly good efficiency to transmissions which are either horizontal or vertical in nature. The helix antenna is said to be circularly polarized. This means it contains both vertical and horizontal portions. Actually, the helix antenna will be most efficient in receiving transmissions which originated from circularly polarized systems. The helix antenna has been used experimentally for uhf and vhf communications and at one time was quite popular with many amateur radio operators who were attempting to build equipment to put them on the microwave frequencies.

All of the alternate antennas discussed in this chapter are presented mainly for academic purposes. Many entire books could be and have been written about the subject of quads, yagis, and helix antennas alone. Those readers who have had experience as amateur radio operators know that it's quite a simple job to put together a half-wave dipole. As the frequency increases, the physical length of the dipole decreases. But at microwave frequencies, other factors become significant and are relatively unknown for antenna designs in the vhf and hf bands. We have already mentioned the critical nature of element dimensions. It becomes very difficult to design many individual elements which have error factors of no more than 1/64th of an inch or even less. Also, different materials used for these elements will necessitate changes in the overall element length. If you use copper wire to make a quad element, for example, its length would be different than if aluminum wire were substituted. At these high frequencies, even the diameter of the wire used will affect the required physical length. Additionally, if you mount your elements on a wooden boom (the section of material which supports the elements) each will have a different length than if an aluminum boom were used. There are many, many variables which must be taken into account and a great deal of experimentation will be required.

During a special antenna test outing that was held for amateur radio operators, I witnessed one ham who had brought a very eleaborate yagi array designed to operate at 1.2 gigahertz. This was not a TVRO antenna by any means, but this fellow's experience will help to properly relay the critical nature of antenna design at these frequencies to the reader. By working out the calculations, this antenna should have produced a gain of about 32 dB. When tests were run at this outing, the first results were unbelievable, so a retest was made. The same results were again obtained, so a third test was conducted. Finally, we had to accept what the tests continued to show us. Although the antenna should have produced a gain of about 32 dB at 1.2 gigahertz, it produced a gain figure of minus 1.3 dB. These gain measurements were in relationship to a single-element dipole. What the negative gain figure indicated was that this amateur operator's elaborate home-built array produced less gain than a single element design. His array consisted of over fifty of these single elements. What happened?

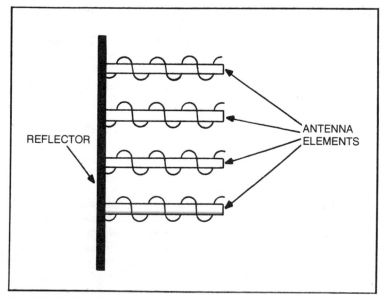

Fig. 5-22. Helix antennas can be combined in parallel to form a helix array.

When the results were made known to the antenna's owner, he did some quick recalculating and realized that the element diameters were figured for a system which used wooden booms for element support. Later, he had decided to use aluminum booms and had not made the proper corrections. A hasty pruning effort after these facts were known brought the antenna close to specs. I believe the final test showed a gain of about 26 dB, which is more in line with the theoretical figure. The amazing part is that the pieces of the antenna elements that were removed during the pruning process were just small slivers. This antenna was designed to operate at 1.2 gigahertz. Each element was about three times the length of the same type of elements which would be used for reception in the TVRO frequency spectrum. If pruning were required of this latter antenna's elements, it might be wiser to use sandpaper to remove fractional bits of metal than to resort to wire clippers. This example points out the highly critical nature of antenna element design.

The reflector antennas discussed earlier in this chapter have only one antenna element instead of fifty. The element is actually a flared piece of waveguide called a feed horn. It is far easier to accurately design a single element than it is to make many, all of which have to be a different physical size. This is one major reason why the dish antennas are so popular at microwave frequencies. The reader is urged to experiment with all types of antenna designs; but if you have not had a lot of antenna experience, especially in the vhf and uhf frequency areas, your move up to microwaves will most likely be a traumatic one.

SUMMARY

It is difficult to say that the antenna is the most important part of a TVRO Earth station. Certainly, without it, no signal will be received at the television set. But this can also be said of the LNA, satellite receiver, modulator, and even the lowly coaxial cables. Each and every part of an Earth station is a necessity, as is each component's efficiency. The largest antenna available coupled with the finest commercial LNAs and receivers will fail miserably if a short length of coaxial cable is defective. TVRO enthusiasts must look at their systems as a whole in order to achieve good results.

The choice of an antenna should be made after considerable research into your personal needs, future wants, the site availability. If you're like most of us, cost will be a prime determining factor as well. You can spend as much as $50,000 or $60,000 for a good commercial antenna (and much more than this without trying too hard), while others will cost less than $1,000. A computer printout of losses in your area will help determine how much gain you need from your antenna; and from that point on, the choice is yours.

If you are interested in building your own antenna, I recommend that you obtain the assembly manual for a commercial kit, of which there are many offered. This will give you a thorough understanding of all that is involved. The manual will probably cost a few dollars, but most manufacturers are happy to supply them, because after all the facts are considered, it is usually about as cheap to purchase the kit than to buy individual components. Perhaps the builder would like to supply some of the parts which are readily available and order the specialty items from the manufacturer. Some companies will probably offer partial kits (mounting kits, azimuth-elevation control kits, feed horn/LNA mount, etc.) which will allow the builder to construct what he can and then complete the project with a few additional kit items. New industries are entering the TVRO field every month, so it will pay to shop around.

Chapter 6
Commercially
Available Antennas

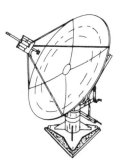

The main distinguishing feature of a satellite Earth station is the huge dish antenna which is bound to draw stares from any passers-by, whether they are electronically inclined or not. The space-age appearance of these devices belies the fact that they are rather simple structures from a purely electronic standpoint. On the other hand, they are highly complex mechanically and their design is quite critical. A deviation in dish curvature of only a fraction of an inch can mean the difference between good reception and poor or nonexistent pictures at the receiving end.

Commercially manufactured TVRO antennas have been available to average consumers for a few years. Each had one thing in common if nothing else. They were all very expensive. Some cost upwards of $10,000, although in recent months, the prices have begun to plummet, with most selling for less than $5,000.

In researching this book, I contacted Wilson Microwave Systems, Inc. in Las Vegas, Nevada. This firm has long been known for its quality antennas designed for many different applications, spanning the spectrum from sophisticated microwave communications to amateur radio antennas.

WILSON MICROWAVE SYSTEMS

Wilson Microwave Systems, Inc. presently offers a 3.35 meter fiberglass dish, dish mounting hardware, manual satellite locator, low-noise amplifier bracket, rotor, and rectangular feed horn for a price of about $1,500. For another $500 or so, you can purchase an extender kit which allows the same dish to be enlarged to a full 4 meters. This kit includes a four-piece fiberglass ring and all mounting hardware. Shown in Fig. 6-1, this antenna can also be had with a remote control option for about $750.

Fig. 6-1. The Wilson 3.35 meter fiberglass dish antenna (courtesy Wilson Microwave Systems, Inc.).

The reason this antenna is so attractively priced is because Wilson leaves the assembly up to the owner, foregoing those expensive labor costs which would be incurred if the same antenna was put together at the factory. This company uses the latest state-of-the-art method of manufacturing the parabolic antenna. The dish is constructed of quarter-inch fiberglass which encloses a full screen mesh. This is the actual reflecting element. The four-piece construction of the dish also provides for easier handling, less installation time, and greatly reduces the shipping costs of this item to the consumer. If you decide at a later time to add the 4-meter extension accessory, it can be easily bolted to the outer edge of the finished 3.35-meter antenna. This kit maintains mechanical rigidity and increases overall antenna performance.

The Wilson design mount for the antenna (called the Vari-Mount) provides extremely easy installation and mounting. This simply requires the digging of four holes with an auger or post hole digger at the antenna site. Each hole is four feet deep and will serve to form a base made up of 2 ×

4s inserted into these holes, which are then filled with about fourteen bags of Redi-Mix.

With the exclusive four-point mount, the owner is assured of a quicker installation and that the antenna will be more securely fastened to the Vari-Mount. The antenna supporting struts aid in stabilizing the fiberglass dish for operational reception in winds of up to 50 to 60 miles per hour. When stability is not adequate in this area, antennas can constantly move back and forth slightly, which is enough to completely interfere with reception.

In order to receive different satellites, a ball-bearing rack is incorporated and allows for easy turning of the antenna. A scale is included on the base to assist in satellite location. The manual satellite locators on the Vari-Mount are in the form of gear-driven hand cranks. They are easily turned and offer a very economical and accurate method of rapidly moving from satellite to satellite.

Wilson offers the optional Satellite Direction Remote Control Console shown in Fig. 6-2, which allows the owner to control the movement of the azimuth and elevation positions of the antenna from the television viewing location. This means indoor operation without having to travel out into the elements whenever another satellite is to be accessed. It is not necessary for this console and the turning levers with their motors to be added upon initial installation. If the hand-controlled design is originally purchased, the remote control option may be added at a later date. Figure 6-3 provides complete antenna specifications on the basic Wilson antenna and on the same antenna with the 4 meter extender. Both are designed for wind survival at speeds of up to 120 miles per hour. Figure 6-4 includes the base specifications as well as azimuth and elevation sweep. Installation time is

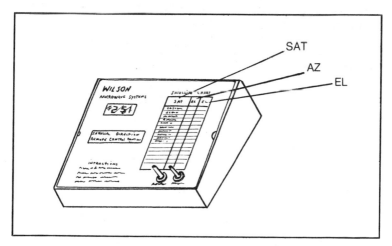

Fig. 6-2. This remote control console allows for azimuth and elevation adjustments from your living room (courtesy Wilson Microwave Systems, Inc.).

ANTENNA	WFD-11		WFD-13
SIZE	11'×¼" (3.35 MTR)		13'×¼" (4 MTR)
CONSTRUCTION	FIBERGLASS	&	SCREEN MESH
FINISH	WHITE		WHITE
WEIGHT	180 LBS		230 LBS
WIND-OPERATIONAL	50 MPH		50 MPH
-SURVIVAL	125 MPH		125 MPH
TEMP RANGE	−60 to +125°F		−60 TO +125°F
FREQUENCY	3.7 TO 4.2 GHZ		3.7 TO 4.2 GHZ
VSWR	1.25 OR LESS		1.25 OR LESS
GAIN	41 DB		42.1 DB
F/D RATIO	.44		.37
1/2 PWR BEAMWIDTH	1.5°		1.5°

Fig. 6-3. Complete specifications for the 3.35 and 4 meter antennas (courtesy Wilson Microwave Systems, Inc.).

approximately 2 hours and requires three persons. The Wilson antenna comes with a one-year warranty.

As was mentioned earlier, the Wilson Microwave Systems TVRO antenna is designed to be assembled at the site. After all parts are received and accounted for, some additional tools and accessories will be needed. A list of the desired extras is shown in Fig. 6-5. The kit contains many parts, and a convenient checklist is included to make sure that one or more components have not been omitted. The complete list of parts is shown in Table 6-1.

Construction involves the selection of a mounting site for the antenna that will permit it to be laid flat on the ground during assembly. It must then be elevated to the operating position with wide clearance from all obstacles. This advice particularly includes power lines, telephone lines, and their associated poles, towers, guy lines, and related structures. The antenna must be firmly anchored to the base, and the exact requirements will depend upon your local soil and weather conditions and any other particulars at the site selected. Wilson strongly recommends the obtaining

BASE

TYPE	AZ/EL
CONSTRUCTION	¼" TO ½" STEEL
WEIGHT	220 LBS
FINISH	GALVANIZED
FOUNDATION	4 - 6" HOLES, 4' DEEP
AZIMUTH SWEEP	80° TOTAL
ELEVATION SWEEP	53° TOTAL

Fig. 6-4. Base specifications for the 3.35 and 4 meter antennas (courtesy Wilson Microwave Systems, Inc.).

1. A PENCIL OR OTHER MARKER.
2. A CARPENTER'S LEVEL.
3. STANDARD BLADE SCREWDRIVERS (LARGE & MEDIUM).
4. A 9/16" WRENCH OF RATCHET WITH 9/16" DEEP SOCKET.
5. (2) 1/2" WRENCHES OR ONE 1/2" WRENCH AND ONE RATCHET WITH 1/2" DEEP SOCKET.
6. A POSTHOLE DIGGER (OR AUGGER) AND SHOVEL.
7. A 1 1/8" WRENCH OR A LARGE CRESCENT WRENCH.
8. A GOOD METAL TAPE MEASURE.
9. COMPASS.
10. 4 STAKES (WOOD OR METAL) AND BALL OF CHALKLINE-ABOUT 50 FEET.
11. 6' STEPLADDER.

Fig. 6-5. Additional tools and accessories that will be needed for installation (courtesy Wilson Microwave Systems, Inc.).

of advice from a local engineer or contractor before starting any work. While these individuals are probably not directly involved with TVRO antenna installations, they will be able to supply you with some sound advice on engineering principles which should be easily understood. In some instances, a surveyor might be needed to provide the proper magnetic headings required to obtain optimum satellite acquisition.

Wilson states that one of the best anchors for the base plate is concrete, with the anchor bolts cast in place. If an alternate method is chosen, you must be certain that it is equally strong and effective. This is most important because in an 8 mile-per-hour wind, a 100-square foot antenna would be subjected to a force of approximately 2,500 pounds. This amount of force can easily upset the antenna should it not be properly mounted. The anchor must be able to withstand this or any greater force that may reasonably be expected to develop. You can probably obtain a record of normal and unusual wind conditions for your particular area from a local radio station or weather recording service.

If the soil in your area is considered to be within normal limits, a reinforced concrete base four feet deep with the anchor bolts cast in the concrete should be adequate. Normal soil, stated by E.I.A. Standard RS-22-C, is "a cohesive type soil with an allowable net vertical bearing capacity of 4000 pounds per square foot and an allowable net horizontal pressure of 400 pounds per square foot per lineal foot of depth to a maximum of 4000 pounds per square foot. Rock, non-cohesive soils, or saturated or submerged soils are not to be considered as normal." If your soil does not meet these requirements, you may need a differently constructed base.

In order to construct a base in a soil which is considered to be within the limits stated above, four holes are dug four feet deep using a conventional post hole digger. Wood forms measuring 2" × 4" are then secured at the tip, and the holes are interconnected with an 8" deep trench, as shown in

Table 6-1. Complete Parts List for the
Wilson Antenna (courtesy Wilson Microwave Systems, Inc.).

PART NUMBER	QUANTITY	DESCRIPTION
WT-803	4	Fiberglass Antenna QTR Panel
WT-804	1	Base Assembly Consisting of WT-811 Fixed Base Weldment and WT-812 Rotating Base Weldment
WT-813	1	Antenna Mounting Frame
AT-100	4	LNA Support Struts, 1" Dia. × 86" Long Alum. Tube
WT-814	1	LNA Mounting Brkt / Tube Weldment
WT-821	4	Antenna Struts, 1" × 1" × 47" Steel Tube
WT-194	4	Anchor Bolts, ¾"-10 UNC × 24" Long
WT-809	2	Actuator
U-100	1	Rotor (Alliance)
WT-832	1	Contour Template Blueprint
-------	1	Hardware Box (see below)
		HARDWARE BOX CONTAINS:
WT-822	2	Elevation Actuator Support Arms, ¼" × 2" × 9⅛" Steel Plate
WT-827	2	Spacer, 3 / 16" × 2" × 3⅜" Steel Plate
WT-823	2	Azimuth Actuator Support Arms ¼" × 2" × 11" Steel Plate
WT-826	2	Spacer, 3 / 16" × 2" × 2½"12 Steel Plate
WT-824	1	Azimuth Actuator Attach Bracket
WT-815	1	Rotor Mounting Plate Weldment, ¼" × 8" × 8" Alum. Plate
FH-1	1	Feed Horn, Rectangular
-------	3	Hardware Bag 1, 2 & 3
		HARDWARE BAG NO. 1 CONTAINS:
B1	4	Ball Bearings, ⅜" Dia.
S49	4	Hex Bolt, ¼"-20 UNC × ⅞" Long
N21	4	Hex Nut, ¼"-20 UNC
N22	4	Lockwasher, Split Ring ¼" Dia.
S24	12	Hex Bolt, 5/16"-18 UNC × 1" Long
S60	4	Hex Bolt, 5/16"-18 UNC × 1½" Long
N01	20	Hex Nuts, 5/16"-18 UNC
N02	16	Lockwasher, 5/16" Dia. Split Ring
S61	8	Flatwasher, 5/16" Dia. Series W
WT-608	24	Hex Bolt, ⅜"-16 UNC × 1½" Long

PART NUMBER	QUANTITY	DESCRIPTION
S56	5	Hex Bolt, ⅜"-16 UNC × 3¾" Long
S66	4	Hex Tap Bolt, ⅜"-16 UNC × 4" Long
WT-090	41	Hex Nut, ⅜"-16 UNC
WT-095	37	Lockwasher, ⅜" Dia. Split Ring
*S63	76	Flatwasher, ⅜" Dia.
S72	4	Flatwasher, 7/16" Dia. Series W
S71	4	Hex Bolt, ½"-13 UNC × 1"
S59	1	Hex Bolt, ½"-13 UNC × 5" Long
S68	2	Hex Bolt, ½"-13 UNC × 1¾" Long
S67	1	Hex Bolt, ½"-13 UNC × 2½" Long
WT-091	4	Hex Nut, ½"-13 UNC
WT-096	4	Lockwasher, ½" Dia. Split Ring
S62	1	Hex Bolt, ⅝"-11 UNC × 9½" Long
S65	1	Hex Bolt, ⅝"-11 UNC × 4½" Long
WT-247	2	Hex Nut, ⅝"-11 UNC
WT-246	2	Lockwasher, ⅝" Dia. Split Ring
N33	1	Turnbuckle, ⅜" Dia.
WT-816	2	Spacers, ⅝" Dia. × .049 × 1⅜" Alum. Tube
WT-831	2	Spacers, ¾" Dia. × .049" × 1½" Alum. Tube
		HARDWARE BAG NO. 2 CONTAINS:
AZ1L	1	Azimuth Scale
CP1L	1	Cover Plate
EL1L	1	Elevation Log Scale
EP1L	1	Elevation Pointer
-----	25 ft.	Vinyl Tape
S70	4	8-32 Machine Screws × ½" Long
CL1L	1	Caution Label
		HARDWARE BAG NO. 3 CONTAINS:
WT-307	8	Hex Nuts, ¾"- 10 UNC

*NOTE: 8 additional washers are supplied in the event that they are needed for contouring the antenna.

Fig. 6-6. Antenna orientation will be discussed later in this chapter, but it is important that this is done properly with regard to the fixed base in order to allow the sweep of the rotating base to acquire all of the satellites of interest.

Preassemble the four anchor bolts to a wooden template with a nut on each side. Allow about ½ inch of the bolts to protrude above the upper nuts, as shown in Fig. 6-6. The template is centered on the form and temporarily nailed into place. The concrete can then be carefully poured into the hole, making sure the reinforcing framework is not disturbed in any way. Once the concrete has been allowed to set the proper amount of time (until it is still workable) the top nuts and the template are removed, making sure that the anchor bolts are not disturbed. It is a good idea to leave the form in place for a 7-day period, since the edges of the concrete will be especially prone to damage during this period of time. Allow the concrete to cure for at least seven days. At this point, the form may be removed by tapping lightly with a hammer to break the form loose. The base can then be installed. This is done by threading a nut two inches down on each of the four anchor bolts. The base is then set on the bolts so that it rests on these four nuts and is secured with four additional nuts on top. Using a carpenter's level, make sure the base is straight and true, making any necessary adjustments by means of the nuts. These can be tightened securely once it has been determined that the base is completely level, as shown in Fig. 6-7.

After the base has been securely fastened down, attach the antenna support frame, as shown in Fig. 6-8. Begin to assemble the antenna by placing three of the quarter panels on a level surface and supporting the centers about 19″ above the surface. The bolts and nuts should be installed loosely at this point. The remaining panel is then added, and all bolts and nuts can be tightened. Note in Fig. 6-9 that the third bolt from the center is not installed at this stage of the assembly. These are installed later when the completed antenna is mounted to the support frame. Also, begin tightening all bolts starting at the outside row and tightening each row sequentially, working inward.

Figure 6-10 shows the proper method of installing the rotor mounting plate to its four support struts. Also shown here is the proper installation of the rotor and low noise amplifier (LNA). Once this step is completed, the assembly is lifted, placed on the antenna, and attached securely.

To install the antenna, lower the mounting frame to a near-horizontal position and support it as shown in Fig. 6-11, or in any equivalent manner. Place the antenna on the four upright ears, and when all holes are aligned, it may be secured with the proper hardware. At this point, all bolts should be snugly secured but not permanently tightened, as it will be necessary to adjust the antenna's contour. First, the antenna support struts and the center turnbuckle are installed, as shown in Fig. 6-12. These elements will be used to adjust the contour.

The secret to obtaining a perfect contour is to have the circumference of the dish perfect to within ⅛ of an inch throughout the whole dish. This is

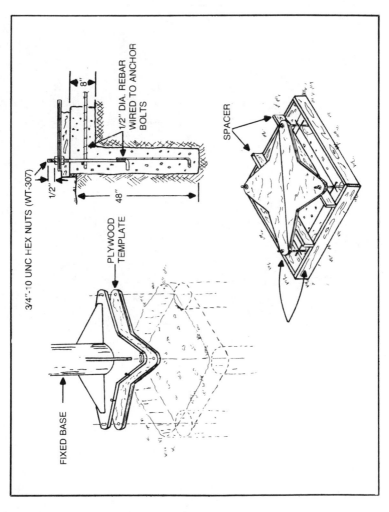

Fig. 6-6. Detailed instructions for the concrete base (courtesy Wilson Microwave Systems, Inc.).

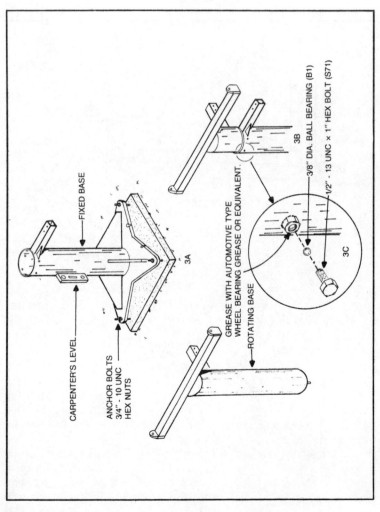

Fig. 6-7. The base is fastened securely once it is measured for level (courtesy Wilson Microwave Systems, Inc.).

CARPENTER'S LEVEL

FIXED BASE

ANCHOR BOLTS
3/4" - 10 UNC
HEX NUTS

3A

3B

GREASE WITH AUTOMOTIVE TYPE
WHEEL BEARING GREASE OR EQUIVALENT.

ROTATING BASE

3/8" DIA. BALL BEARING (B1)

1/2" - 13 UNC x 1" HEX BOLT (S71)

3C

94

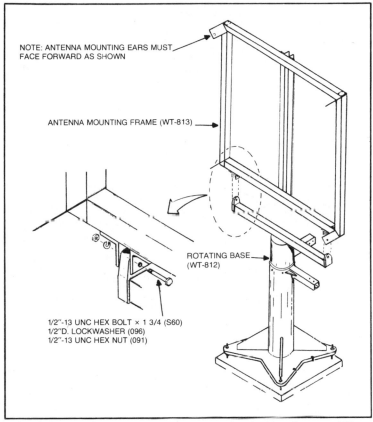

NOTE: ANTENNA MOUNTING EARS MUST FACE FORWARD AS SHOWN

ANTENNA MOUNTING FRAME (WT-813)

ROTATING BASE (WT-812)

1/2″-13 UNC HEX BOLT × 1 3/4 (S60)
1/2″D. LOCKWASHER (096)
1/2″-13 UNC HEX NUT (091)

Fig. 6-8. The antenna support frame is attached to the base (courtesy Wilson Microwave Systems, Inc.).

almost an impossible feat in the normal manufacturing process when four separate panels are involved. However, Wilson has overcome this problem by making the panels slightly smaller and making up the difference with the aid of washer spacers, thus allowing the installer to vary the circumference as necessary.

In order to adjust the antenna's contour to obtain a perfect parabolic shape, three different adjusting systems are utilized: spacers, support struts, and the center turnbuckle. If the circumference is proper, the turnbuckle should be in the neutral position (no force applying pressure either up or down) on the dish center. The turnbuckle may, however, be used for fine adjustment of the contour.

The first step in contour adjustment involves making a contour template which must be perfectly accurate. Wilson includes with their antenna complete instructions on how to make this template, and even states that the large box that the fiberglass quarter panels arrived in will

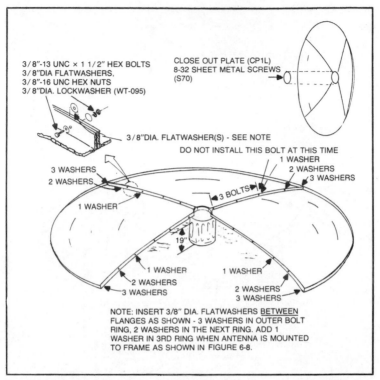

Fig. 6-9. All bolts are tightened sequentially working from the outside of each section (courtesy Wilson Microwave Systems, Inc.).

suit quite adequately. The reason that the template must be absolutely accurate is that even an error of ¼″ will cause a considerable loss of gain. Once you are satisfied that the template is perfectly accurate, and with the antenna installed on the antenna support frame in the horizontal position, the template is placed in the dish, as shown in Fig. 6-13. The contour is checked across each panel seam and at 45° from the seams as shown in Fig. 6-14.

Figures 6-15, 6-16, and 6-17 show some of the variations that may be noted when checking the contour of your dish. Figure 6-15 provides the correct contour. If this is obtained with the template placed across both seams of the dish, no further adjustments are necessary. However, if a perfect contour is obtained in one position, with the contour pictured in Fig. 6-16 obtained across the other seams, it will be necessary to adjust the antenna support struts outward. It should now be apparent why none of the securing hardware was tightly bolted in place when the antenna was mounted on its base.

Once the antenna support struts have been adjusted, the contour should be checked once again. You may find that the contour in one direction has been corrected, while the opposite contour had been changed to a slight degree. If the change is minor, the correction may be made by tightening the center turnbuckle to add tension, thereby pulling down the center and returning to the perfect contour. If, however, the change is somewhat pronounced, it will be necessary to remove one washer from each of the first and second holes at all four seams, referring back to Fig. 6-12 for placement of this hardware.

The contour variation in Fig. 6-16 indicates that the circumference is too large. It is now necessary to make the circumference smaller by removing one washer at the first and second holes at all four points. It is important to note here that this particular contour will have the highest

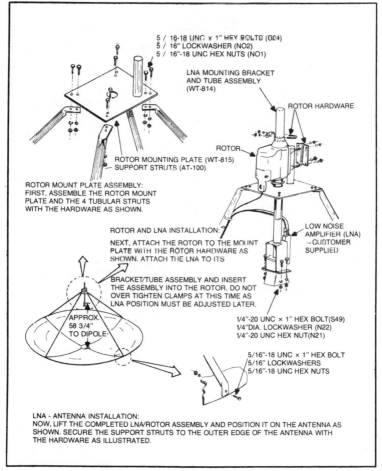

Fig. 6-10. Installation of the rotor mounting plate (courtesy Wilson Microwave Systems, Inc.).

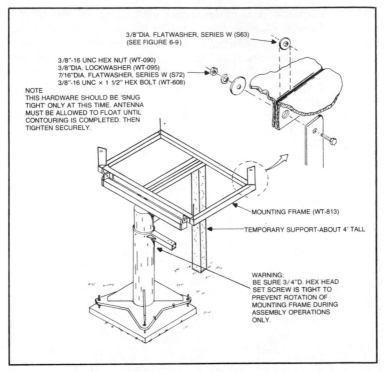

3/8"DIA. FLATWASHER, SERIES W (S63)
(SEE FIGURE 6-9)

3/8"-16 UNC HEX NUT (WT-090)
3/8"DIA. LOCKWASHER (WT-095)
7/16"DIA. FLATWASHER, SERIES W (S72)
3/8"-16 UNC × 1 1/2" HEX BOLT (WT-608)

NOTE
THIS HARDWARE SHOULD BE 'SNUG TIGHT' ONLY AT THIS TIME. ANTENNA MUST BE ALLOWED TO FLOAT UNTIL CONTOURING IS COMPLETED. THEN TIGHTEN SECURELY.

MOUNTING FRAME (WT-813)

TEMPORARY SUPPORT-ABOUT 4' TALL

WARNING:
BE SURE 3/4"D. HEX HEAD SET SCREW IS TIGHT TO PREVENT ROTATION OF MOUNTING FRAME DURING ASSEMBLY OPERATIONS ONLY.

Fig. 6-11. The antenna is installed by lowering the frame as shown here (courtesy Wilson Microwave Systems, Inc.).

amount of loss, since errors at the tips of the dish cause more gain loss than errors at the center of the dish.

Figure 6-17 shows yet another type of variation that may be encountered when checking the contour of the dish. Here, the dish has a circumference that is too small. In order to correct this condition, it is necessary to add one washer at the first and second holes at all four points.

Once the contour has been adjusted properly, cover the flange seams with the vinyl tape and the antenna can be raised sufficiently to install the elevation and azimuth actuators. This is shown in Fig. 6-18. A temporary support should be used for this purpose. The first step is to install the elevation actuator support arms and spacer and the azimuth actuator attach brackets with the proper hardware. The hex nuts are not tightened until the studs on the elevation actuator have been inserted into their mounting holes. The ½″ diameter holes in the antenna support frame angles which best suit your particular satellite elevation requirements should be chosen. The upper end of the actuator is then attached. The installation of the azimuth actuator will follow this same basic procedure. The rod-end is secured between the attach brackets. Also, be sure to loosen the set screw on the azimuth before any azimuth rotation is attempted.

Figure 6-19 shows the proper method for installing the azimuth and elevation scales, as well as a caution label. Satellite azimuth and elevation logging scales are provided so that once a particular satellite's position has been obtained, it's location can be recorded. This provides a method of easy access for relocating the satellite at a later date.

To install the elevation scale, peel off the protective backing from the label and press it into the elevation hinge ear. The pointer label is placed directly above it on the hinge angle with the pointer on or near the scale. This is accomplished by folding the lower edge around the edge of the hinge angle. Now, locate the azimuth scale on the top edge of the fixed base tube with the "0-degrees" mark under the pointer and the azimuth actuator in its compressed position.

The coax and rotor cable are now installed following proper instructions as provided by Wilson. After all connections to the receiver and the

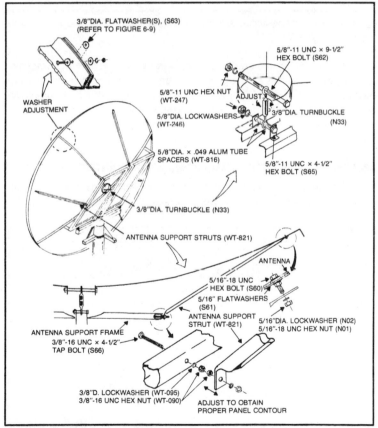

Fig. 6-12. Installation of the antenna support struts and center turnbuckle (courtesy Wilson Microwave Systems, Inc.).

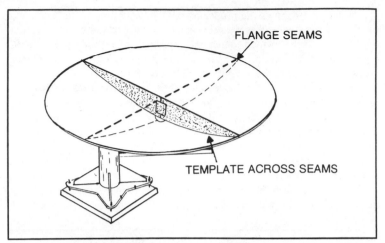

Fig. 6-13. The perfectly accurate template is placed in the dish (courtesy Wilson Microwave Systems, Inc.).

TV set have been completed, orient the antenna to the satellite you have chosen and adjust the LNA by sliding it up and down until a clear, sharp picture is obtained. At this point, installation is complete and all clamps can be tightened securely. Figure 6-20 illustrates what the completed antenna will look like once it has been installed and wired to receive satellite transmissions.

MICROWAVE ASSOCIATES COMMUNICATIONS TVRO ANTENNAS

Microwave Associates Communications is a company which is involved quite heavily in manufacturing products for commercial Earth sta-

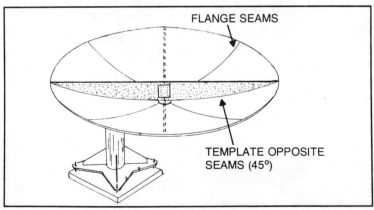

Fig. 6-14. The contour is checked at the points indicated here (courtesy Wilson Microwave Systems, Inc.).

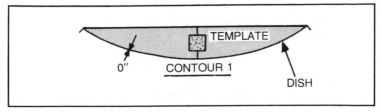

Fig. 6-15. If this contour is obtained, no further adjustments are necessary (courtesy Wilson Microwave Systems, Inc.).

tions. This company's products extend from the lowest to the highest end of the electromagnetic spectrum—from dc to visible light. They offer three TVRO antennas which may be of interest to persons who intend to install their own Earth station.

Figure 6-21 shows the 3-meter antenna, which is economical and intended for applications where signal strength is adequate. The antenna is built in eight sections, with a unique integrated backup structure for installation on a rigid hour angle mount. This antenna is available with a single or dual polarized feed system.

Figure 6-22 shows a top and side view of the antenna and its mounting structure. The foundation for this mount consists of three footings formed in auger-drilled holes. Typically, 0.6 cubic yards of concrete will be used in each of the holes. Figure 6-23 provides complete specifications for this antenna, which offers a gain of 40.1 dB at 5 GHz.

Figure 6-24 shows the larger 3.7-meter antenna from Microwave Associates Communications. The larger dish is designed for applications where signal strength of the satellite transmissions is moderate. The antenna is built in four sections and also includes an integrated backup structure for installation on the hour angle mount. The mounting site is prepared in a similar manner described for the previous 3-meter antenna. Figure 6-25 provides the specifications for the 3.7-meter antenna, which offers a higher gain than the previous model.

For areas with weak satellite signals, the 4.6-meter might be considered. It offers a 44 dB gain and is built in two sections of rigid molded

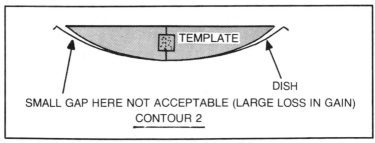

Fig. 6-16. If the contour appears like this, the circumference is too large (courtesy Wilson Microwave Systems, Inc.).

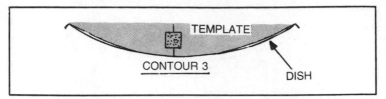

Fig. 6-17. If this variation occurs, the circumference is too small (courtesy Wilson Microwave Systems, Inc.).

fiberglass for easy installation on a circular track mount. The foundation for this mount is a reinforced slab typically ten feet square with seven to ten cubic yards of concrete, depending upon the soil-bearing capacity. Figure 6-26 shows a side and top view of the antenna and associated mounting structure. This antenna is also available with a receive/transmit feed which allows it to be used for transmitting as well as receiving purposes. Figure 6-27 provides complete specifications.

The antennas just described are actually designed for commercial cable television applications and are built to more rigid specifications than are some of the antennas designed specifically for personal Earth stations. The price of these antennas will reflect the commercial nature of their design.

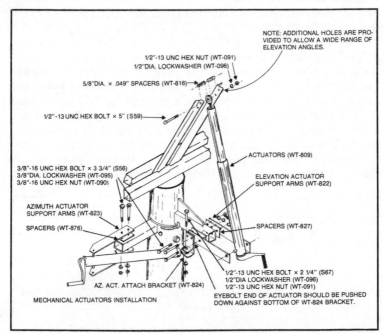

Fig. 6-18. Installation of the elevation and azimuth actuators (courtesy Wilson Microwave Systems, Inc.).

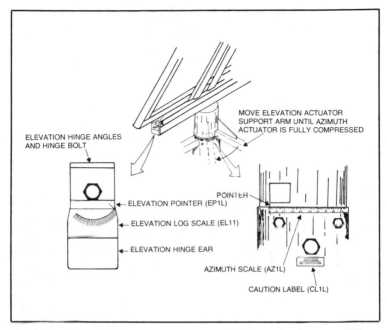

ELEVATION HINGE ANGLES
AND HINGE BOLT

MOVE ELEVATION ACTUATOR
SUPPORT ARM UNTIL AZIMUTH
ACTUATOR IS FULLY COMPRESSED

POINTER

ELEVATION POINTER (EP1L)

ELEVATION LOG SCALE (EL11)

ELEVATION HINGE EAR

AZIMUTH SCALE (AZ1L)

CAUTION LABEL (CL1L)

Fig. 6-19. Installation of the azimuth and elevation scales (courtesy Wilson Microwave Systems, Inc.).

SKYVIEW I ANTENNA

The Skyview I spherical antenna from Downlink, Inc. in Putnam, Connecticut, is a high-performance, low-cost spherical antenna design which allows signal reception from several satellites by moving the feed horn assembly only. This means that the smallest portion of the antenna is adjusted as to its physical relationship with the spherical dish, which remains stationary. The antenna is constructed of rugged angle iron framework, durable mahogany wood lattice, and is covered with a woven aluminum screen for maximum reflection of signals and low wind loading characteristics. Shown in Fig. 6-28, this antenna is delivered unassembled, leaving the labor up to the owner. A fair amount of savings can be had by putting it together yourself. Geometrically, the antenna is a 12' by 12' section from the surface of a sphere with a 60' diameter. In other words, it can be thought of as a section cut from a large ball.

This is a shallow antenna with a center depth of only 7.25 inches and a long focal length of fifteen feet. The curvature of the reflector closely approximates the curvature of a shallow parabolic reflector with the same focal length and exhibits similar characteristics in concentrating the signals to a precise focal point.

Like the parabolic antenna, the spherical type focuses the intercepted microwave energy by reflecting it from its surface to a point in front of the

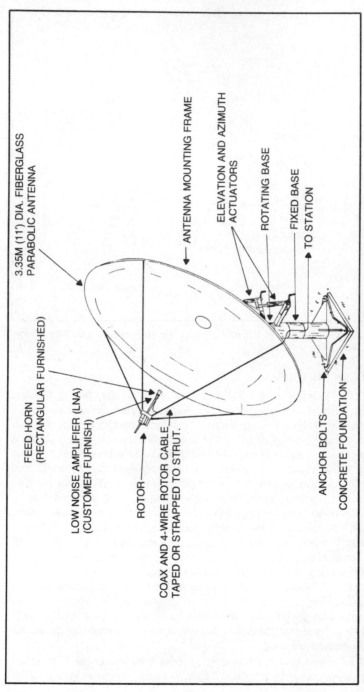

FEED HORN
(RECTANGULAR FURNISHED)

3.35M (11') DIA. FIBERGLASS
PARABOLIC ANTENNA

ANTENNA MOUNTING FRAME

ELEVATION AND AZIMUTH
ACTUATORS

ROTATING BASE

FIXED BASE

TO STATION

LOW NOISE AMPLIFIER (LNA)
(CUSTOMER FURNISH)

ROTOR

COAX AND 4-WIRE ROTOR CABLE
TAPED OR STRAPPED TO STRUT.

ANCHOR BOLTS

CONCRETE FOUNDATION

Fig. 6-20. This is what the completed Wilson antenna will look like (courtesy Wilson Microwave Systems, Inc.).

Fig. 6-21. The Microwave Associates 3-meter antenna (courtesy Microwave Associates Communications).

dish. The parabolic antenna has one focal point directly in front and perpendicular to the center of the antenna. For this reason, it must be aimed as an entire unit directly at the satellite whose signal is to be received. Unlike parabolic designs, the spherical antenna focuses microwave signals received as much as plus or minus 20 degrees off of the line which is perpendicular to the center. It does this with a negligible loss of efficiency. This means that a spherical antenna positioned to reflect and focus the signal from one satellite will simultaneously reflect the signals of several adjacent satellites and focus them at different points in front of the antenna. A feed horn properly positioned at one of these focal points will amplify the maximum amount of signal that this antenna is capable of reflecting from a given satellite. This assumes that all satellite signals are received by the antenna at less than a 20 degree angle from the center line. By moving the feed horn and LNA combination which is not mounted directly to the dish,

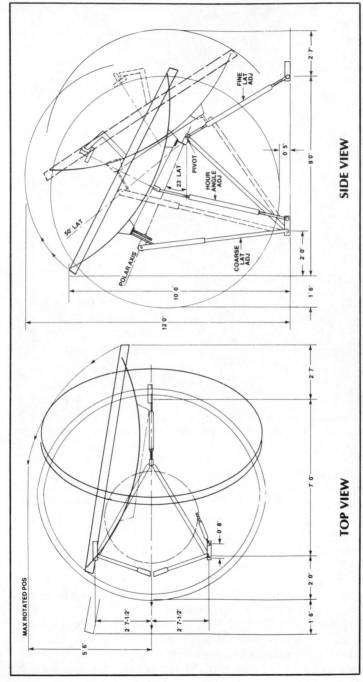

SIDE VIEW

FINE LAT ADJ

50° LAT

23° LAT

PIVOT

HOUR ANGLE ADJ

POLAR AXIS

COARSE LAT ADJ

2' 7"

0' 5"

9' 0"

2' 0"

1' 6"

10' 0"

12' 0"

TOP VIEW

MAX ROTATED POS

0' 8"

2' 7"

7' 0"

2' 0"

1' 6"

2' 7-1/2"

2' 7-1/2"

5' 6"

Fig. 6-22. Shown here is both a top and side view of the 3-meter antenna (courtesy Microwave Associates Communications).

SPECIFICATIONS

GENERAL

Model Number	3700-3
Operating Frequency Range	3.7 to 4.2 GHz
Feed Type	prime focus
Mid-Band Gain (Ref to OMT Port)	40.1 dBi @ 4 GHz
VSWR	1.2:1 maximum
Polarization	single or dual linear
Feed Port Isolation	30 dB minimum
Half-Power Beamwidth	1.7° nominal
First Sidelobe Level	17 dB typical
Radiation Pattern, Averaged Sidelobe Envelope per FCC 25-209	
32-25 log θ	$1° \leq \theta \leq 48°$
10 dBi	$48° \leq \theta \leq 180°$
Antenna Noise Temperature, °K (Referenced to OMT Port)	
Elevation 5°	50
10°	40
20°	30
30°	26
40°	25
Feed Flange	CPR-229G
Mount Configuration	polar axis

ENVIRONMENTAL

Temperature Range	
Operational	−40 to +55°C
Wind Loading @ 0°C(+32°F)	
Air Temperature	
Operational	96 km/h (60 mi/h) wind, gusting to 136 km/h (85 mi/h)
Survival	200 km/h (125 mi/h) with no ice; 140 km/h (87 mi/h) with 51 mm (2″) radial ice
Atmospheric Conditions	salt, pollutants, and corrosive contaminants as encountered in coastal and industrial areas
Shipping Weight	440 lbs. (200 kg)
Shipping Volume	75 Ft.^{3} (23 m^3)

Fig. 6-23. Complete specifications for the 3-meter antenna (courtesy Microwave Associates Communications).

Fig. 6-24. Microwave Associates' 3.7 meter antenna (courtesy Microwave Associates Communications).

as in the parabolic design, signals can be received from several adjacent satellites without repositioning the spherical dish. A further discussion of spherical antennas is included in another chapter.

The Skyview I antenna, with its screened reflective surface, provides very good wind loading characteristics and will be unaffected in mild winds of less than 40 miles per hour. Think of it as a screen door. Air currents pass through the tiny holes and do not affect the overall structure as much as would be the case if a solid surface were provided. If a solid sheet of metal were used to replace the screen on the antenna, then wind loading would be severe. Wind loading can be thought of as the amount of pressure exerted on any surface by air currents.

At wind velocities of over 50 miles per hour, the screened surface of the Skyview I antenna begins to react similar to that of a solid surfaced

SPECIFICATIONS

GENERAL

Model Number	3700-3.7
Operating Frequency Range	3.7 to 4.2 GHz
Feed Type	prime focus
Mid-Band Gain (Ref to OMT Port)	41.7 dBi @ 4 GHz
VSWR	1.2:1 maximum
Polarization	single or dual linear
Feed Port Isolation	30 dB minimum
Half-Power Beamwidth	1.5° nominal
First Sidelobe Level	18 dB typical
Radiation Pattern, Averaged Sidelobe Envelope per FCC 25-209	
$32-25 \log \theta$	$1° \leqslant \theta \leqslant 48°$
10 dBi	$48° \leqslant \theta \leqslant 180°$
Antenna Noise Temperature, °K (Referenced to OMT Port)	
Elevation 5°	47
10°	38
20°	28
30°	24
40°	23
Feed Flange	CPR-229G
Mount Configuration	polar axis (hour angle)

ENVIRONMENTAL

Temperature Range	
Operational	−40 to +55°C
Wind Loading @ 0°C (+32°F) Air Temperature	
Operational	96 km/h (60 mi/h) wind, gusting to 136 km/h (85 mi/h)
Survival	200 km/h (125 mi/h) with no ice; 140 km/h (87 mi/h) with 51mm (2″) radial ice
Atmospheric Conditions	salt, pollutants, and corrosive contaminants as encountered in coastal and industrial areas
Shipping Weight	690 lbs. (314 kg)
Shipping Volume	240 Ft.^{3} (74m^3)

Fig. 6-25. Specifications for the 3.7 meter antenna (courtesy Microwave Associates Communications).

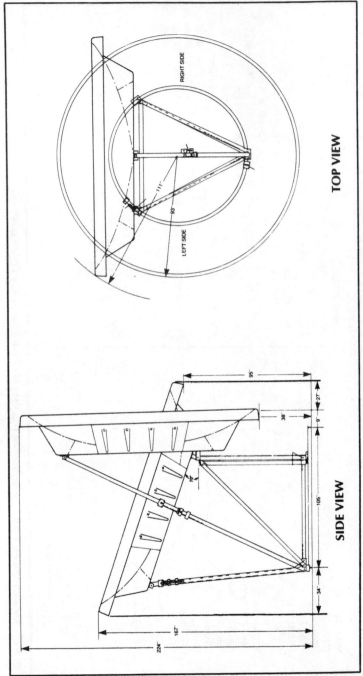

Fig. 6-26. Side and top view of the 4.6 meter antenna, along with its mounting structure (courtesy Microwave Associates Communications).

SPECIFICATIONS

GENERAL

Model Number	3700-4.6
Operating Frequency Range	
Receive	3.7 to 4.2 GHz
Transmit (Optional)	5.925 to 6.425 GHz
Feed Type	prime focus
Mid-Band Gain	
(Ref to OMT Port)	
Receive	44 dBi @ 4 GHz
Transmit (Optional)	46.5 dBi @ 6.15 GHz
VSWR	1.2:1 maximum
Polarization	single or dual linear
Feed Port Isolation	30 dB minimum
Half-Power Beamwidth	
Receive	1.2° nominal
Transmit (Optional)	0.79° nominal
First Sidelobe Level	18 dB typical
Radiation Pattern, Averaged Sidelobe	
Envelope per FCC 25-209	
32 - 25 log θ	1° ≤ θ ≤ 48°
10 dBi	48° ≤ θ ≤ 180°
Antenna Noise Temperature, °K	
(Referenced to OMT Port)	
Elevation 5°	44
10°	37
20°	28
30°	24
40°	22
Feed Flange	CPR-229G
Mount Configuration (Circular El / Az Track)	
Elevation Range	5 to 70°
Azimuth Range	360°

ENVIRONMENTAL

Temperature Range	
Operational	−40 to +55°C
Wind Loading @ 0°C(+32°F)	
Air Temperature	
Operational	96 km/h (60 mi/h) wind, gusting to 136 km/h (85 mi/h)
Survival	200 km/h (125 mi/h) with no ice; 140 km/h (87 mi/h) with 51 mm (2") radial ice
Atmospheric Conditions	salt, pollutants, and corrosive contaminants as encountered in coastal and industrial areas
Shipping Weight	2950 lbs. (1341 kg)
Shipping Volume	609 Ft.³ (186m³)

Fig. 6-27. Specifications for the 4.6 meter antenna (courtesy Microwave Associates Communications).

111

design, and the antenna will be subjected to a very great force in high winds. For this reason, it is essential to firmly anchor the antenna to the ground with the sturdy angle iron framework included. When properly installed, the antenna will withstand winds of up to 125 miles per hour.

As is the case with most antennas, the exact requirements of the foundation will depend upon weather conditions, soil conditions, and other criteria applicable at the site. As was mentioned previously, the Skyview I antenna comes as a kit of parts and you perform all of the assembly. This procedure is performed in four stages: angle iron frame assembly; wood lattice assembly; screening; and rear support assembly. Downlink advises that while moving the completed antenna, care must be exercised to prevent bending and twisting which may adversely affect the screened surface. Once the completed antenna is mounted to its supporting structure, the entire assembly will be quite rigid and durable.

In order to give the reader an idea of what is involved in the actual home construction of this antenna, the following information has been gleaned from the assembly manual provided by Downlink. Any home assembly project is heavily dependent upon building instructions provided by the manufacturer. Even the finest kit on the market will be a nightmare to the builder if the assembly manual is not easy to understand, accurate, and filled with pictures and drawings. In this area, Downlink, Inc. has excelled. The manual is straightforward, listing each minute step separately. There are pictures and/or charts on almost every page. This should serve to facilitate home construction as well as to teach the builder every possible aspect of the project to make routine servicing at a later date much easier.

The first major stage of construction is to assemble the angle iron frame. This starts with the horizontal rib bracing. The ribs maintain the spherical shape of the antenna and care must be taken to insure the accuracy of the given measurements. Each completed rib must be solidly braced. The assembly of the horizontal rib bracing is completed in six steps for each rib. After step six is completed, you go back to step one and start on the second rib. There are a total of three horizontal ribs in this antenna.

Next comes the vertical rib assembly. The vertical ribs are marked for location of static tuning points and then attached to the horizontal ribs to make up the main portion of the angle iron framework. There are five vertical ribs. This assembly is completed in four easy steps for each of the five vertical ribs. The angle iron framework is completed when the frame reinforcement steps are accomplished. This is finished in four steps. The finished angle iron framework will look similar to Fig. 6-29.

Part two of construction involves the wood lattice assembly. The first phase is preparing the five vertical sections of the lattice structure for mounting. The completed strips are then marked for proper positioning of the horizontal lattice strips and to indicate the exact measuring points for static tuning. This procedure is accomplished in six steps. Next comes the vertical lattice assembly, where the five vertical strips are fitted with

112

Fig. 6-28. The Skyview I spherical antenna (courtesy Downlink, Inc.).

hardware and mounted to the angle iron frame. It takes four steps to complete this section.

Phase three is the construction of the horizontal lattice assembly, where thirteen strips of the horizontal lattice are nailed into their proper positions and then secured with brass screws to the vertical lattice strips. Six steps complete this phase.

Phase four involves the corner assembly, where the corner portion of the wood lattice is constructed prior to the attachment of the horizontal end strips. The corners add rigidity and strength to the lattice and provide the even surface necessary for the proper screening.

Phase five completes the wood lattice by attaching the horizontal end strips and providing an even edge for attaching the screen. This is accomplished in seven steps.

The final phase is called static tuning, which is the curvature of the wood lattice to the approximate spherical shape of the dish. This curvature is shown in Fig. 6-30. A complete set of measurements is provided by Downlink. When this step is completed, the structure should appear similar to that shown in Fig. 6-31.

The third major step in construction is screening. The reflective surface of this antenna consists of aluminum mesh screen, commonly referred to as woven hardware cloth. When carefully laid on the pre-curved wood lattice, it makes an excellent spherical reflector for microwave signals in the 3.7 to 4.2 GHz range.

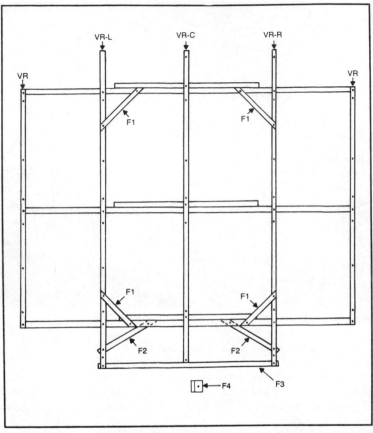

Fig. 6-29. The completed angle iron framework of the Skyview I (courtesy Downlink, Inc.).

The screen is rolled onto the wood lattice, as shown in Fig. 6-32, and stapled into place. This assembly must be performed carefully and with patience in order to obtain the smoothest surface possible. This is mandatory because a consistent surface, free of wrinkles, bubbles and indentations, must be provided to assure maximum gain from the antenna. Screening is accomplished in seven steps.

The final major step is the building of the rear support assembly, which will allow the completed antenna to be raised and fitted into place. This major step is divided into two phases, the first being the rear leg support assembly, completed in four steps, and the second, the assembly of the rear support structure, done in five steps.

After the dish portion of the antenna is fully completed (at the end of the fourth major step), antenna mounting and tuning is started. This involves the use of a template and is easily done from the information

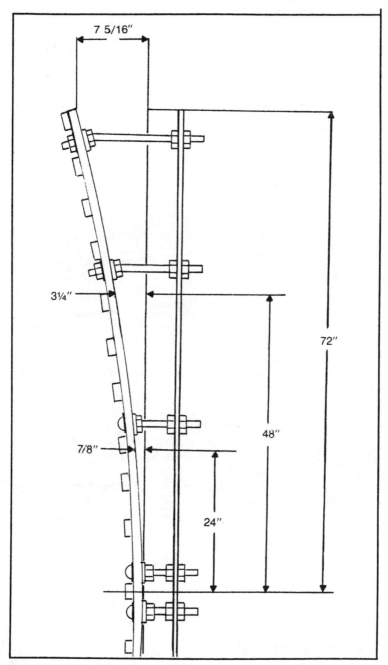

Fig. 6-30. Tuning is done to match the curvature of the wood lattice to the shape of the dish (courtesy Downlink, Inc.).

115

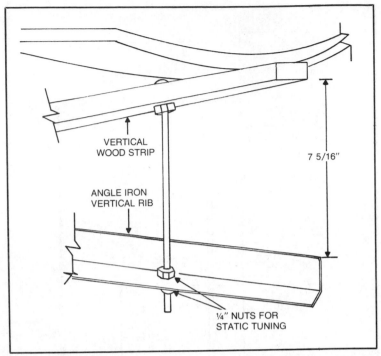

VERTICAL
WOOD STRIP

ANGLE IRON
VERTICAL RIB

7 5/16"

¼" NUTS FOR
STATIC TUNING

Fig. 6-31. When static tuning is completed, the structure should appear like this (courtesy Downlink, Inc.).

provided in the instruction manual. Next, the LNA/feed horn combination is mounted. This includes the feed horn, rotor and control, LNA bracket, motor bracket and hex key. A mounting stand is then built for the feed horn and connected to a concrete base. Figure 6-33 shows the completed feed horn assembly in its final mounting position.

The Skyview I antenna is available in three different models. The one discussed here is the 12-foot size, but 10 and 8 foot designs are also offered. Figure 6-34 provides specifications for all three models, which are basically the same except for physical size and gain ratings. These antennas should provide many years of useful service with minor maintenance. The angle iron frame is coated with a rustproof primer and may be repainted as desired. The wood lattice and screen may also be painted, as this will not affect the antenna's performance. Downlink suggests that snow should be brushed off the antenna after a storm, as this definitely can affect picture quality. Also, a continual buildup of snow can damage the mesh screen.

WINEGARD EARTH STATION ANTENNAS

Winegard has long been a company associated with the construction of television antennas. Most of these have been vhf and uhf yagi designs for

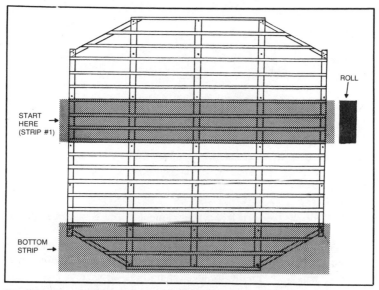

Fig. 6-32. The screen must be carefully and patiently rolled onto the wood lattice (courtesy Downlink, Inc.).

standard reception. This company has now entered the TVRO Earth station field by offering two antenna packages. They also offer the complete Earth station system, including LNAs, receivers, feed horns, etc.

The Winegard 3-meter antennas use a compression-molded eight-segment fiberglass reflector and are supplied complete with prime-focus feed and polar (hour-angle) mount. The LNA is supported at the focal point of the antenna in four gigahertz versions. This antenna is particularly easy to transport due to its compact shipping dimensions and can be readily erected on its foundation in about three hours.

The Winegard 10-foot parabolic dish contains reflector segments which are field-replaceable and interchangeable. This is good to know should one panel become damaged due to wind or other conditions. Here,

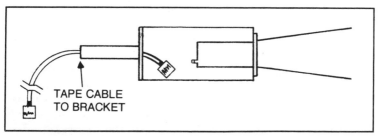

Fig. 6-33. The completed feed horn assembly in its final mounting position (courtesy Downlink, Inc.).

117

```
                          SPECIFICATIONS

Size                12'                    10'                    8'
Equivalent (Meters) 3.66                   3                      2.44
Frequency           3.7-4.2 GHz            3.7-4.2 GHz            3.7-4.2 GHz
Construction        Angle Iron,            Angle Iron,            Angle Iron,
                    Mahogany,and           Mahogany,and           Mahogany, and
                    Woven Hardware Cloth   Woven Hardware Cloth   Woven Hardware Cloth
Weight              266 lbs.               245 lbs.               199 lbs.
Wind-Operational    75 mph                 75 mph                 75 mph
Wind-Survival       125 mph                125 mph                125 mph
Gain                40.5 dB                39.5 dB                37.6 dB
```

Fig. 6-34. Complete specifications for the Skyview 8, 10, and 12 foot antennas (courtesy Downlink, Inc.).

all that would be required is the replacement of the damaged segment, which is far less expensive than replacing the entire dish.

The 10-foot dish is rated at a midband gain of 40.1 dB and at a maximum vswr of 1.2:1. The feed implemented on Winegard Earth station antennas is the prime focus type with the low noise amplifier located at the focal point. The feed horn assembly is designed to mount into the molded-in recess collar of the reflector. Mounting stability is accomplished by compression clamping the flange collar of the reflector between the feed mount plate and the associated retaining clips.

In their literature, Winegard points out that your choice of antenna size will depend upon your exact location in the 48 contiguous states. While a 10-foot dish is ideal in Kansas City, for instance, a 12-foot dish is recommended for a southern Texas location. The 12-foot dish offered by Winegard uses a 10-segment compression molded reflector and is supplied with the same polar mount. At any location within the 48 contiguous states, this mount allows realignment to any satellite located between 70° and 140° west longitude without foundation changes.

The 12-foot parabolic dish can be installed on its foundation in less than four hours and is shipped in a small enough package that it can be transported to the installation via a pickup truck. This antenna offers a midband gain of 41.7 dB.

SATVISION MODEL SV-11

Shown in Fig. 6-35, the SatVision SV-11 has a diameter of 11 feet and is assembled at the site. A rotatable feed is standard with this dish and allows the LNA to be rotated 90° for reception of both vertical and horizontally polarized satellite signals. The antenna is assembled on a flat surface and involves the installation of several reflector sections, feed horn braces and then the tightening of all bolts. SatVision also offers a complete line of Earth station components, as well as complete Earth station packages. This includes antennas, receivers, LNAs, and even big screen color television sets. This company produces an excellent introductory brochure describing

118

the basic TVRO field. For further information, write SatVision International, P. O. Box 1490, Miami, Oklahoma, 74354.

COMTECH 3-METER ANTENNA

Shown in Fig. 6-36, the Comtech 3-Meter Satellite Earth Station Antenna is designed to provide economical high performance for reception from current and projected geostationary satellites. The Comtech 3-Meter Earth Station Antenna offers a high-performance guarantee and excellent cost-effective operation. This operation, coupled with versatility, ease of transportation, quick installation, and minimum maintenance, makes it a popular design.

Each antenna consists of a precision-molded, three-piece parabolic reflector surface, a dual-axis mount, and its own feed system. The whole assembly can be easily installed by two men in approximately two hours. No special tools, panel alignments, or testing is required to achieve maximum sensitivity from this antenna. The three-piece configuration allows for very economical transportation and facilitates an uncomplicated

Fig. 6-35. The SatVision model SV-11 antenna (courtesy SatVision International).

119

Fig. 6-36. The Comtech 3-meter satellite Earth station antenna (courtesy Comtech Antenna Corp.).

assembly in remote sites or on rooftops where handling equipment is not readily available. Incidentally, this reflector is also available in a one-piece configuration for bulk trailer shipment.

The newly-designed EL/AZ mount provides a sturdy support for the antenna system and requires only simple adjustments for accessing any satellite. The Comtech-developed joint lines on the three-piece reflector assure minimum surface distortion and maximum rigidity.

Additional features include a fiberglass reflector and steel mount, easy installation of single or dual polarized linear feed systems, and excellent stability. Figure 6-37 provides complete specifications.

COMTECH 5-METER ANTENNA

For marginal signal areas, the additional gain provided by a 5-meter antenna is a most welcome addition to any TVRO Earth station. Shown in Fig. 6-38, the Comtech 5-Meter Antenna Assembly consists of a three-piece fiberglass parabolic reflector and a prime focus feed system with an elevation over azimuth adjustable pedestal. Construction is of a high-strength, corrosion-resistant fiberglass. The mounting holes for the structure are drilled with precision machining fixtures to facilitate accurate and trouble-free assembly on site.

The linearly-polarized, high-efficiency feed is located at the prime focus of the parabolic reflector. A three-step Tchebyscheff transformer provides optimum impedance transformation from the radiator to the input flange. The feed assembly can be rotated manually through 360° in order to change or to peak polarization.

The pedestal, in its basic configuration, is a tripod structure with clamping devices at the base of each leg. Full azimuth coverage (0 to 360°) is provided by an eight-foot diameter ring at the base of the pedestal. Elevation (0 to 60°) is accomplished through translation of the rear support tube.

The 6 meter antenna assembly requires only a limited amount of maintenance due to the lack of electronic devices and moving parts. No special tools, test equipment or materials are required to maintain this high-quality system. A visual inspection every six months for antenna security and tightness will be sufficient to insure unlimited operation.

An option is provided which utilizes a sub-reflector with a surface tolerance of 0.020 inches. Called the Cassegrain Option, this permits the

General Description

Reflector Type	Parabolic, 3 piece fiberglass construction
Reflector Diameter	3 meters
Mount Configuration	El / AZ
Arc Coverage	360° in azimuth and 0 to 60° in elevation
Foundation	Drilled pier type or suitable roof mount

Electrical

Operating Frequency Range	3.7 - 4.2 GHz
Gain	40.3 dB @ 4.0 GHz
VSWR	1.3:1 max.
Polarization	Single or dual linear, manually adjustable
Cross-polarization Isolation (single polarization)	35 dB min.
Isolation Between Ports (dual polarization)	30 dB min.
Half power beamwidth	1.7° nominal @4.0 GHz
Feed Interface	CPR—229F flange

Environmental

Wind Loading

Operational	60 MPH (100 KMPH)
Survival	100 MPH (160 KMPH)

Temperature Range

Operational Wind	−5° to +130°F (−20° to +55°C)
Survival	−30° to +130°F (−35° to +55°C)
Pointing Accuracy	0.25° in 60 MPH wind
Atmospheric Conditions	Salt, pollutants, and corrosive contaminants as encountered in coastal and industrial areas

Weight (approx.)

Shipping	550 lb.
Net	450 lb.
Shipping Volume	135 cubic ft.

Fig. 6-37. Specifications for the 3-meter antenna (courtesy Comtech Antenna Corp.).

incorporation of various speed systems at a point easily accessible during operation as well as at the time of installation.

While certainly applicable to personal TVRO Earth stations, the size and cost of this 5-meter antenna mean that it is a design which is marketed to CATV operation. Figure 6-39 gives the specifications for this design. Total cost for the basic antenna is $5,775.

Comtech also offers a 4-meter Earth station antenna which is nearly identical with the earlier 3-meter design already discussed in this chapter. It offers an additional two dB of gain at four gigahertz. Present basic price of the three-meter antenna is $2,500, while the four-meter design will cost about $1,000 more.

SUMMARY

There are many antennas which are available to TVRO Earth station enthusiasts through commercial channels. In looking over the large selection, it should be noted that many are designed for cable company installations. These will generally be larger and somewhat more rugged than those

Fig. 6-38. The Comtech 5-meter antenna (courtesy Comtech Antenna Corp.).

```
Mechanical
Reflector Diameter ............................................................................5 meters (16 feet)
Mount Type ...............................................................................Elevation / Azimuth
Reflector Surface Tolerance ..................................................................0.060″ RMS static
Survival Wind Loads ...............................................90 MPH winds with ½″ radial ice
                      ...................................................................... 125 MPH winds—no ice
Survival Temperature ................................ −40° to 160°F (−40° to 70°C)
Arc Coverage .............................................................................Elevation 0° to 60°
                      .................................................................................... Azimuth 360°
Net Weight .............................................................................1500 lbs. approx.
Shipping Weight ......................................................................1800 lbs. approx.
Shipping Cube ..........................................................................600 cu. ft.
Reflector Finishes .....................................................................White, standard
Operating Conditions ...................................Salt, pollutants, and corrosive
                      ...................................... contaminants as encountered in coastal
                      .................................................................. and industrial areas

Electrical
Frequency ...............................................................................3.7 — 4.2 GHz
                      ................................................................................. 5.925 to 6.425 GHz
Polarization ............................................................................Single or dual linear,
                      ........................................................................ manually adjustable
Gain @ 4 GHz .........................................................................44.3 ± 0.2 dB
       @ 6 GHz .........................................................................47.3 + 0.2 dB
Half Power Beamwidth @ 4 GHz .................................................1.1°
                      @ 6 GHz ..............................................................0.7°
VSWR ....................................................................................1.3 max.
Input Flange ............................................................................CPR—229F
Options Available
Cassegrain Feed
Dual Linear Polarization
Circular Polarization
```

Fig. 6-39. Complete specifications for the 5-meter antenna (courtesy Comtech Antenna Corp.).

intended specifically for personal Earth station use. They also cost substantially more. Commercial uses of TVRO stations are far more critical than necessary for most personal installations, so these may not fall into a practical realm for many persons who are interested only in an Earth station for their backyards. In some fringe areas, it may become necessary to resort to a commercial dish antenna for this wider diameter in order to produce a stronger signal at the LNA input.

Most personal Earth station antennas measure ten to thirteen feet in diameter. Many of the ten-foot antennas will accept a dish extension kit which enlarges the overall diameter to about four meters.

In satellite receiving stations, the parabolic dish antenna will often be the most expensive component. Make your choice wisely, but do not scrimp on dish surface area simply to save money. The manufacturer you choose to supply your antenna will be able to recommend the least expensive package which will provide good operation in your area. Some locations will make excellent use of the smaller ten-foot size, while others will require a full four meters or even more in rare instances for the same signal strength and quality of reception at the television set.

If you live in an area which is subjected to high winds on a seasonal basis, you may wish to discuss this with the antenna manufacturer. Knowing your circumstances, the supplying company may wish to provide you with a more rugged antenna mount which will assure dependable reception regardless of weather conditions.

Chapter 7
Commercially Available
LNAs, Receivers,
and Modulators

The precision and complexity involved in the electronic equipment used to receive the satellite transmissions and convert them into signals which are usable by your present television receiver are not as applicable to home construction as are other types of home entertainment equipment. A later chapter does include some information on the few kits available to the home builder; but for now, most Earth station owners will use commercially available equipment which is offered by many manufacturers.

This chapter will discuss some of the products which are presently offered. Some companies may market a single piece of equipment such as a receiver, LNA, or modulator; but most offer complete stations comprised of all of these components plus an antenna and feed line as a package. It will be necessary in most cases for the owner to install a base on which to mount the antenna and to set up the electronic equipment by making the proper cable attachments, but no actual construction of electronic circuits is required.

The following manfacturers and their products are known throughout the TVRO Earth station field and are representative of the market as a whole. Prices will vary in direct accordance with the complexity of the equipment, its efficiency, and the many features offered.

As a general rule of thumb, the basic TVRO Earth station components can be purchased in the following price ranges. A commercial low noise amplifier will usually cost between $700 and $1,100. The actual price will depend upon noise factor, gain, and installation criteria. Most satellite receivers will cost between $1,000 and $3,000, with about $1,500 being the average. The less expensive receivers are often not equipped with remote tuning facilities. We are speaking here of receivers designed for the personal Earth station. Commercial equipment designed for cable television operation can easily cost two or three times the figures quoted. Modulators are the least expensive major components in the Earth station and may be had for between $100 and $200. Some receivers offer built-in modulator circuits and are priced accordingly.

The reader is urged to absorb the information presented in this book and then to contact several manufacturers for up-to-date pricing and additional information. From the specification sheets provided in this chapter, you should be able to get an idea of what is applicable to your personal situation. From this point on, it should be easy to narrow the field of choices, finally opting for the equipment that is most ideal for your uses.

The low noise amplifier is mounted directly to the feed horn of the antenna and serves to increase the strength of the satellite signal at this point. The amplified signals are then fed into a short length of coaxial cable, the other end of which is attached to the satellite receiver. Low noise amplifiers are essential in all TVRO earth stations. Even with the largest antenna, you must still have an LNA to receive anything at all. Low noise amplifiers generally carry two ratings which are important to the buyer. The first is noise temperature, which is rated in degrees Kelvin. An LNA with a 120° noise temperature has a noise figure of 1.5 dB. This means that, in simple terms, the LNA circuitry will add 1.5 dB of noise to the signal it receives from the antenna. This is a very low figure, but noise is detrimental to any receiving system and must be kept to especially low values in Earth stations. This 120° figure is representative of the great majority of LNAs used for satellite television reception. Some areas, however, which do not enjoy strong satellite signals may require even better noise figures of their LNAs. A chart is provided in the appendices, along with a general signal strength map of the United States, which will help you in determining what noise figure your LNA must provide. As noise temperature drops, the noise figure decreases. A 100° LNA introduces less than 1.3 dB of noise to the receive system, while an 80° rating is indicative of approximately 1.05 dB of noise. Obviously, the lower noise figures provide better reception, but LNAs in this category cost far more than those rated at 120° K. Fortunately, here in the United States, a 120° unit should be adequate for most personal users, although in some areas, the 100° units are to be preferred. The charts in the appendices are designed for commercial users such as cable television systems. Personal Earth stations can get by with a slightly higher noise figure. The manufacturer who supplies your LNA will be in a good position to advise you as to the specifics for your area.

AVANTEK AWC - 4200 SERIES

The Avantek LNA is used exclusively by many firms who offer complete TVRO Earth station packages. Shown in Fig. 7-1, this device has a maximum noise figure of 120° K. (1.5 dB) and a gain of 50 to 60 dB, depending upon the model chosen. The Avantek Model 4215 seems to be the one most often chosen for personal TVRO Earth stations.

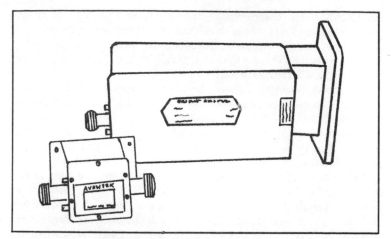
Fig. 7-1. The Avantek 4200 series low noise amplifier (courtesy Avantek, Inc.).

These low noise amplifiers offer excellent performance and reliability as stand-alone preamplifiers for receivers in the 3.7 to 4.2 gigahertz band, which encompasses all of the satellite TV frequencies. They combine extremely low noise figures with relatively high output powers. This allows the Earth station owner to include multipliers or filters ahead of the microwave receiver without cutting signal strength to the receiver input to a marginal value.

Each amplifier is powered by low-voltage dc. Power is applied through the rf output connector by means of the coaxial feed line. This arrangement simplifies antenna feed point mounting, since a single cable carries both dc and rf power. This may also provide added protection against low noise amplifier damage due to lightning-induced transient voltages when compared to conventional unshielded power cabling which is sometimes used without a grounded conduit.

The Avantek low noise amplifiers are packaged in aluminum cases with an integral waveguide input flange and a coaxial cable connector output. The cases are coated with a urethane finish and the lids are sealed with elastomeric 0-rings to assure that the critical amplifier circuitry is completely watertight when mounted in an unprotected environment.

For commercial purposes, these amplifiers are excellent for antenna feed point mounting when used as the sole preamplifier in light and medium capacity Earth terminals, placing them ahead of all feed line losses for best system noise performance. The low 1.5 dB makes them very suitable for almost every personal Earth terminal application. Specifications are given in Fig. 7-2.

A large number of Avantek low noise amplifiers are currently in service in small Earth terminals in commercial and personal installations. This company has worked with users of commercial and military downlink

GUARANTEED SPECIFICATIONS @ 25°C

Model	Frequency Range (GHz) Min	Gain (dB) Min/Max	Noise Figure (dB) Max	Noise Temp. (°K) Max	Gain Flatness (dB) Max	Gain Slope (dB/MHz) Max	Linear Group Delay (ns/MHz) Max
AWC-4205	3.7–4.2	47/53	1.5	120	±.5	.01	.01
AWC-4206	3.7–4.2	57/63	1.5	120	±.5	.01	.01

Model	Parabolic Group Delay (ns/MHz^2) Max	Ripple* Component (ns Peak-to-Peak) Max	AM to PM Conv. deg/dB at −5 dBm Output Max	Power Output at 1 dB Comp. Point (dBm) Min	Intercept Point for third order Intermod Products (dBm) Min	VSWR Maximum In	VSWR Maximum Out	Input Power Nominal VDC	Current (mA) Typical
AWC-4205	.001	0.1	0.1	+10	+20	1.25	1.5	+ 15 to + 28	150
AWC-4206	.001	0.1	0.1	+10	+20	1.25	1.5	+ 15 to + 28	170

*The Ripple Component is defined as the residual group delay obtained after equalization of the linear and parabolic components by the "chord" method in any 40 MHz band.

Fig. 7-2. Specifications for the 4200 series LNAs (courtesy Avantek, Inc.).

low noise amplifiers, and their engineers have an excellent background of experience in the Earth terminal field. Again, most major suppliers of Earth station terminals either offer this manufacturer's low noise amplifiers as standard equipment or as options.

Figure 7-3 shows the basic dimensions of the 4200 series low noise amplifier. The input is provided with a ferrite isolator to assure a minimum vswr without compromising noise figure performance. Put in layman's terms, this means that the low noise amplifier provides a good match of impedance between the antenna and the transmission line. Gallium arsenide field effect transistors (GaAs FETs) are used in the input stages and are combined with bipolar intermediate and output stages for an optimum combination of low noise figure, dynamic range, and power capability.

An integral monolithic IC voltage regulator permits these amplifiers to operate with an input voltage anywhere in the +15 to +28 Vdc range. This effectively isolates all of the amplifier stages from variation in the dc supply voltage level and from noise or hum appearing on the supply line. The regulator contains internal thermal overload protection, short-circuit current limiting, and output transistor safe area compensation to protect the expensive GaAs FETs from damage. Protection against reverse polarity is also provided should you connect the power supply incorrectly.

Once the low noise amplifier is installed, it should require no further attention throughout the operating lifetime of the TVRO Earth station. The manufacturer states that there is no measurable time-related performance degradation. Additionally, the amplifier does not required periodic adjustment.

Dependability of these LNAs is enhanced by the use of extremely reliable Avantek-designed and fabricated GaAs FETs and silicon bipolar transistors. These components have documented histories of proven performance in some of the most critical microwave applications including airborne and marine communications, radar, and even equipment aboard orbiting satellites.

For special applications, Avantek produces a series of 120° LNAs that may be powered from low voltage dc, 115 or 230 V power lines, and from ac mains with an automatic changeover feature that will switch to a standby dc supply in case of a power failure. These optional features are far more applicable to commercial installations than to the personal Earth station user. If you have a power failure at your home, chances are your television receiver won't be working anyway. Extremely low noise temperature versions of the LNAs described are also available on special order. These are rated at 95° Kelvin, which breaks down to a noise figure of less than 1.25 dB.

Avantek Inc. is located at 3175 Bowers Avenue, Santa Clara, California, 95051, but people see their products sold through many outlets and by many manufacturers of TVRO Earth station packages. One of my favorite sources for electronic equipment is Long's Electronics, P.O. Box 11347, Birmingham, Alabama. This mail order company has been offering amateur

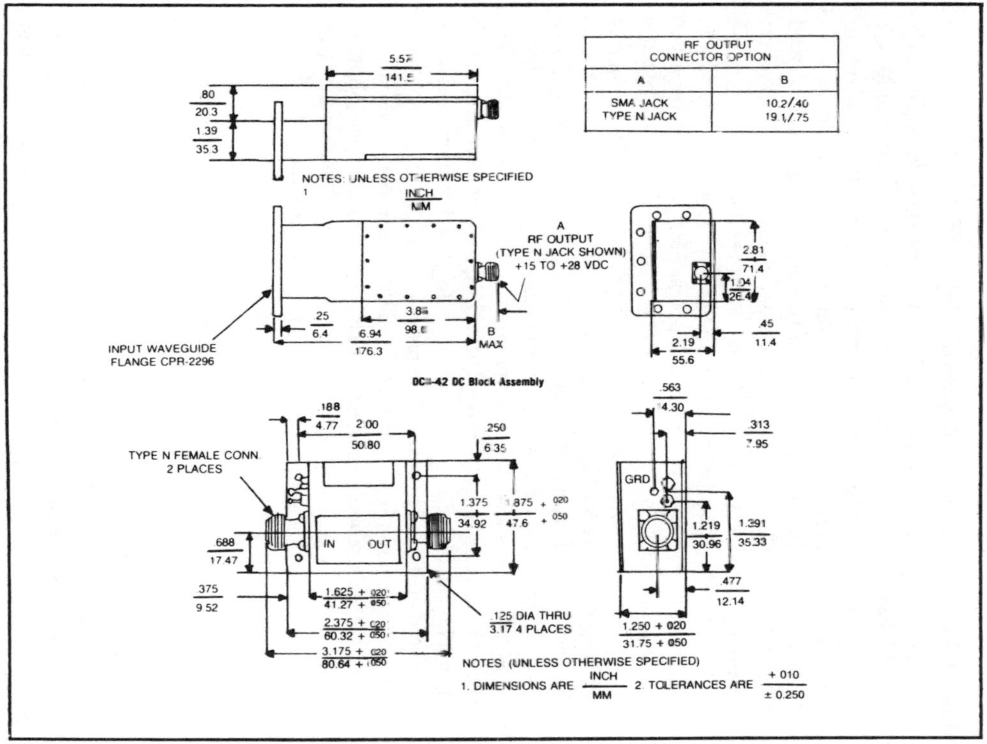

Fig. 7-3. Block diagram and dimensions of the 4200 series LNA (courtesy Avantek, Inc.).

radio equipment at discount prices for many years and they are now marketing a number of TVRO Earth stations. This is not meant to be a commercial for this outfit, but they are one of the few companies who are offering the products of many different Earth station manufacturers instead of producing their own. The Avantek 4215 low noise amplifier is being offered by Long's as of this writing for about $750. This price may not be in effect by the time this book is available but is indicative of discount pricing in the latter part of 1980.

The Avantek 4215 mentioned here is part of the 4200 series just discussed. Its specific gain of 50 dB is typical and it requires about 150 milliamperes of dc current at a potential of 15 to 28 Vdc for normal operation.

MICROWAVE ASSOCIATES COMMUNICATIONS COMPANY LNA

Microwave Associates Communications Company offers a broad line of TVRO Earth station components, most of which are designed for commercial applications. Their low noise amplifier is shown in Fig. 7-4. It is available in four different models, all of which offer a minimum gain of 50 dB. The difference in the models is found in their noise figures. Each model uses gallium arsenide field-effect transistors and is equipped with surge suppression circuitry and a rugged weatherproof container.

Each model is designed to operate from a dc voltage of +15 to +28 Vdc and at a maximum current of 125 milliamperes. They will operate properly over a wide ambient temperature range of −50 to +60° centigrade and are completely waterproof. Figure 7-5 shows the dimensions of this amplifier, which is equipped with an input flange for mating to the antenna feed horn and a coaxial output connector. Figure 7-6 gives the specifications for this series, with breakdowns on the noise figures of the four models offered.

Microwave Associates Communications can be contacted at 63 Third Avenue, Burlington, Mass., 01803. The LNA discussed here is also being marketed by manufacturers of complete TVRO Earth stations and often it may carry the label of the latter company.

The reader may find several other LNAs which carry different manufacturers' labels. Close examination of the specifications will show that these may actually be made by the two companies listed here and then sold by these other companies. As the TVRO field becomes larger and larger, more manufacturers who are presently offering their LNAs only in the commercial field will begin to sell to the personal market. The two manufacturers of LNAs discussed in this chapter were chosen because they supply their products to a large number of other companies.

AVANTEK LNA/DOWNCONVERTER

When the LNA-amplified signal arrives at the satellite receiver, the microwave frequency is converted to a lower value by the receiver's front end. It is this frequency which the receiver demodulates in order to supply

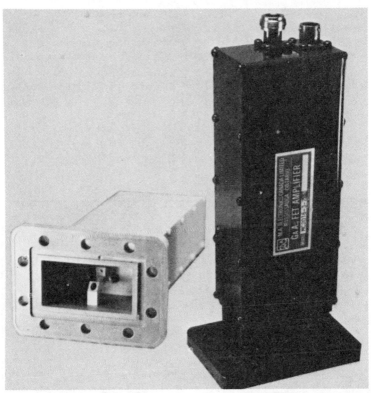

Fig. 7-4. The Microwave Associates low noise amplifier (courtesy Microwave Associates Communications).

its video and audio output. In coaxial cable, higher frequency transmissions are attenuated far more than are those at lower frequencies. In the TVRO industry today, there is an increasing trend to build this conversion circuitry into the output of the LNA. Here, the amplification process takes place, the microwave frequencies are downconverted to a lower frequency, and then the output is fed to the receiver by the coaxial cable. This process has simply taken part of the receiver circuitry and placed it at the opposite end of the transmission line near the LNA. When this is done, there is less loss in the coaxial cable.

The Avantek ACA-4220 shown in Fig. 7-7 is an excellent example of this circuit. It performs block downconversion at the antenna feed point. All power is provided through the coaxial feed line. Basically, this circuit is a combination of the earlier LNA discussed and a downconverter circuit all built into a single case which is mounted at the antenna feed point.

The ACA-4220 is a completely self-contaned assembly of low-noise GaAs FET preamplifier, block downconverter, local oscillator, filter, and i-f amplifier. This is essentially a complete antenna-mounted receiver front-

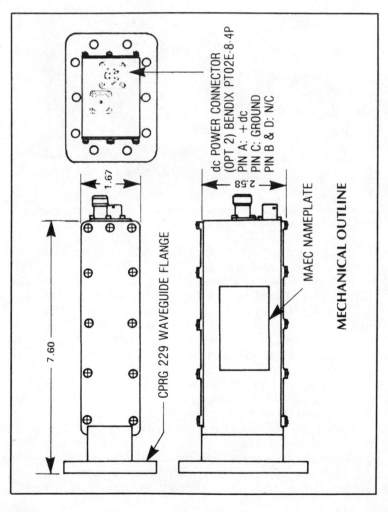

Fig. 7-5. Mechanical dimensions for this LNA. (Courtesy Microwave Associates Communications).

dc POWER CONNECTOR
(OPT 2) BENDIX PT02E-8-4P
PIN A: +dc
PIN C: GROUND
PIN B & D: N/C

MAEC NAMEPLATE

MECHANICAL OUTLINE

CPRG 229 WAVEGUIDE FLANGE

1.67

7.60

2.58

SPECIFICATIONS

GENERAL

Model Number	MC2016
Frequency Band	3.7 to 4.2 GHz

Noise Temperature / Noise Figure Options @ 20°C

Part Number	Noise Temperature	Noise Figure
1808567-4	85°K	1.2 dB
1808567-2	100°K	1.3 dB
1808567-3	110°K	1.4 dB
1808567-1	120°K	1.5 dB

Noise Figure Variation (−50 to +60°C)	0.01 dB V°C typical
Gain @ 20°C	50 dB minimum
Gain Variation (−50 to +60°C)	±2.7 dB maximum
Frequency Response (3.7 to 4.2 GHz)	±0.5 dB
Amplitude Gain Slope	0.015 dB / MHz maximum
Gain Stability	±0.5 dB / day maximum
AM-to-PM Conversion	0.1%/dB maximum to −5 dBm output power
VSWR (In / Out)	1.20/1.5 maximum

Output Power

@ Third Order Intercept	+20 dBm minimum
@ 1 dB Gain Compression Point	+10 dBm minimum
Group Delay	< 0.001 ns / MHz maximum
	< 0.001 ns / MHz$_2$ maximum

PRIMARY POWER

Voltage	+15 to +28 Vdc
Current	125 mA maximum

ENVIRONMENTAL

Operating Temperature	−50 to +60°C
Relative Humidity	100%

MECHANICAL

Input Flange	CPR-229G
Input Connector	Type N Female
Dc Power Connector*	Bendix PT02E-8-4P**
Length	8 inches maximum
Weight	2 lbs.
Waveguide Pressurization	10 psig maximum

*Coax center conductor or solder feedthrough available on special order.
**Mating connector PT06E-8-4S (SR) supplied.
All specifications are subject to change without notice.

Fig. 7-6. Specifications for Microwave Associates LNAs (courtesy Microwave Associates Communications).

133

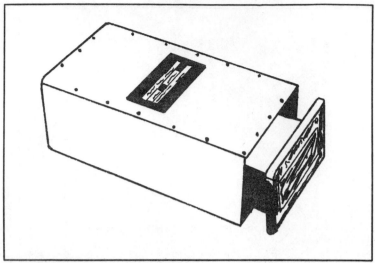

Fig. 7-7. The Avantek ACA-4220 LNA/downconverter (courtesy Avantek, Inc.).

end for TVRO Earth stations. Converting the entire 3.7 to 4.2 gigahertz Downlink band to a convenient 1440-940 megahertz intermediate frequency range, one of these units provides a critical front-end performance for virtually any number of receivers. The 940-1440 megahertz i-f range was selected specifically because it offers the minimum possibility of interference from other services. This unit features a minimum overall gain of 57 dB and a 120° overall equivalent noise temperature. Specifications are given in Fig. 7-8. The LNA/downconverter is not appreciably larger than a standard low noise amplifier. Its dimensions are shown in Fig. 7-9. A later chapter in this book on home building of TVRO components will explain downconverters in more detail.

Remember, the two main ratings which you should be interested in when purchasing an LNA are noise (rated in degrees Kelvin) and gain (rated in decibels). I am not aware of any manufactured LNAs which offer noise temperatures higher than 120°. In most locations, this will be the minimum noise figure that is tolerable for good reception and when using standard antennas. Lowest acceptable gain should be at least 45 dB. Amplifiers which don't meet these specifications may be completely useless, so use the section of this chapter with the manufacturers' specifications to make comparisons with other amplifiers you may see advertised. As long as the ratings compare favorably with those of the commercial units discussed here, other LNAs should be very suitable.

After the satellite signal is gathered by the antenna, amplified by the LNA, and passed down the transmission line, it enters the receiver. Here, a specific satellite channel is selected and the audio and video information is detected. The receiver is the most complex component (electronically) in

Guaranteed Specifications, −40 to +50°C (−40 to +122°F)

Input Frequency, MHz:	3700-4200	Linear Group Delay, ns/MHz (Note 1):	0.01 max.
Output Frequency, MHz:	940-1440	Parabolic Group Delay, ns/MHz^2 (Note 1):	0.001 max.
LO Stability, MHz:	±6 max.	Group Delay, Ripple, ns Peak-to-Peak (Note 1):	0.1 max.
LO Frequency, MHz:	5140 or 5160 (with option), 5140 (without option)	Spurious Outputs, dBc (with B_0 average equal to −15 dBm):	−50 min.
Optional LO Frequency MHz	+15 to +20 VDC, LO=5·40	LO Radiation, dBm (at input or output):	−45 max.
Shift Switching:	+22 to +28 VDC, LO=5·160	Frequency Response of Input Port, (Note 2) dB down (relative to passband response):	−40 min. (DC-3 GHz, 5-61 GHz, 6.6-10 GHz)
Single Sideband Noise Figure, dB:	1.5 max.		
Gain, dB:	63 max., 55 min.		
Gain Variation vs. Frequency dB (entire band):	±1 dB max.		
Gain Slope, dB/MHz:	0.01 max.		
Image Response (6.1-6-6 GHz) Relative to Passband Response, dB:	−60 min.	Input VSWR (50 Ω):	1.25:1 max.
		Output VSWR (50 Ω):	2:1 max.
Typical Output Intercept Point for Intermodulation Products, dBm:	+10	Power Supply Voltage, volts:	+15 to +28 VDC (see optional LO shift)
		Current, mA:	150 typ.
Phase Noise, dBc min. relative to 200 KHz deviation measured (3 KHz BW):	−58 at 10 KHz decreasing to −65 at 200 KHz, −66 dB, 200 KHz to 4.2 MHz	Operating Temperature Range:	−40° to +50°C (−40° to +122°F)
		Operating Humidity Range:	0 to 100% Relative Humidity

Note 1: Linear and Parabolic components of best fit equation to the Group Delay Curve in any 40 MHz band by the "Chord" method. Ripple Component is the residual group delay obtained after equalization of the linear and parabolic components.

Fig. 7-8. Specifications for the 4220 LNA/downconverter (courtesy Avantek, Inc.).

135

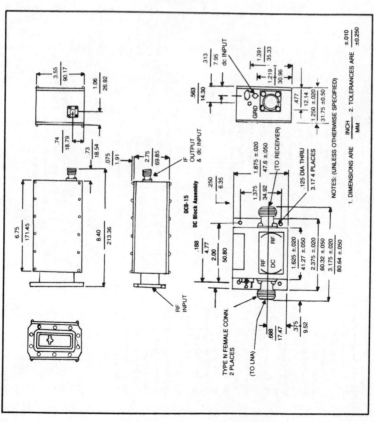

Fig. 7-9. Block diagram and dimensions of the 4220 LNA/downconverter (courtesy Avantek, Inc.).

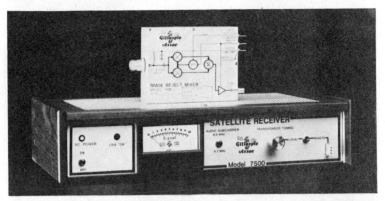

Fig. 7-10. The Gillaspie model 7500 receiver (courtesy Gillaspie & Associates).

the entire TVRO Earth station, but it is far simpler than your color television set. However, its design is more critical and far less tolerant of circuit faults. Some receivers are offered which must be mounted at the television set in order to allow for the changing of satellite channels. Others are designed to be mounted at the antenna and are enclosed in waterproof cases. The antenna-mounted receiver is to be preferred in most applications, because this will decrease the needed length for the feed line from the LNA. Longer feed lines produce greater losses. TVRO Earth stations must deal with signals from space that arrive at the reception site at extremely small power levels. The TVRO Earth station owner must do all that is in his power to minimize these losses. The following discussion has incorporated manufacturers' information about their satellite receivers. The manufacturers chosen are known to be reputable and their products are currently in use at TVRO Earth stations within the United States and elsewhere. Again, the manufacturer is in the best position to advise you as to the applicability of their products to your personal needs. This discussion will put you in the ball park when choosing a satellite receiver. Additional manufacturer information will help with the final decision as to which model to buy.

GILLASPIE & ASSOCIATES MODEL 7500

The Model 7500 satellite receiver from Gillaspie & Associates, 177 Webster Street, Monterey, California, 93940, is shown in Fig. 7-10. It is designed to receive signals in the 3.7 to 4.2 gigahertz satellite television band and is available with an optional remote control console, which is shown in Fig. 7-11A.

The receiver comes in two parts, the main section and the downconverter. The latter mounts at the antenna. The Model 7500 provides two independent audio channels at 6.8 and 6.2 megahertz. Other audio outputs may be obtained at different frequencies as an option. The circuitry uses

Fig. 7-11A. Gillaspie's receiver is available with this optional remote control console (courtesy Gillaspie & Associates).

automatic frequency control to insure stability and prevent drift. An LNA theft alarm is also included as part of the receiver circuitry and is activated should the LNA be removed from the antenna.

Unusual in most satellite TV receivers, the Model 7500 includes a signal strength carrier level meter on the front panel. A built-in crystal controlled video monitor may be included with this receiver. This provides an output which is directly usable by your home television set. There is a built-in video polarity inversion switch and another special circuit is included which is designed to eliminate signal flutter.

The optional remote control console offers 24-position detented transponder selection, continuous fine tuning, and easily accessible tuning controls in the rear. The remote control is recommended in situations where there is a large distance between the television receiver and the ground station.

The image reject mixer (downconverter) is mounted at the antenna and uses low-cost RG59-U cable to span the distance to the receiver. This makes longer antenna-to-receiver runs possible. Input frequency is 3.7 to 4.2 gigahertz at 50 ohms. The output frequency is 70 megahertz at 75 ohms. Typical noise figure of this stage is 13 dB, while image rejection is 15 dB typical. Conversion gain is rated at 18 dB typical.

The image reject mixer unit is normally mounted directly on the low noise amplifier or very close to it, usually within six feet. These two major units encompass the entire receiver system. The current price of the basic

Model 7500 receiver system is around $1,500. Figure 7-11B shows a graph of image rejection, noise figure, and conversion gain curves over the satellite television frequencies.

VITALINK COMMUNICATIONS CORPORATION MODEL V-100

The Vitalink Multiple Receiver System consists of a Model V-100C block downconverter and model V-100 video receiver. The system has 24-channel capability and is intended primarily for multiple dwelling applications such as apartments, condominiums, and mobile home estates. The V-100C is a high-stability downconverter that accepts a cable connection for a 4 gigahertz antenna system. Power for the low noise amplifier is inserted on the coaxial cable from the V-100C. This feature may be disabled at will. The converter brings the entire 12-channel group from either vertical or horizontal polarization down to a uhf frequency. This channel group is further amplified and split, providing four ports for connection to the Model V-100R receiver.

The V-100R is a broadcast-quality video receiver that features completely automatic fine tuning and gain control under pushbutton panel selection. Each V-100R has four standard audio subcarrier outputs and also a built-in modulator for vhf channels 3 or 4. The receiver is shown in Fig. 7-12, while the downconverter is shown in Fig. 7-13.

The Vitalink Multiple Receiver System has a threshold of 8 dB carrier-to-noise (C/N) performance and is designed with solid-state technology to provide maximum flexibility and desired features at a low cost for commercial-grade equipment.

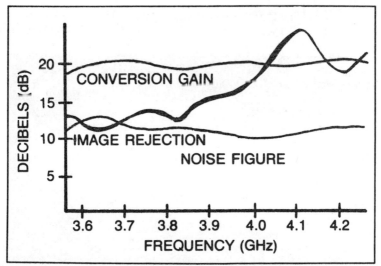

Fig. 7-11B. This graph shows image rejection, noise figure and conversion gain curves (courtesy Gillaspie & Associates).

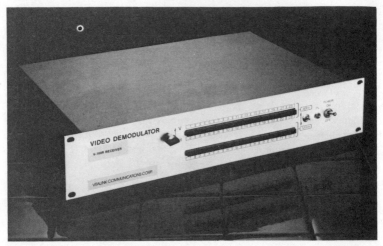

Fig. 7-12. The Vitalink V-100 video receiver (courtesy Vitalink Communications Corp.).

Figure 7-14 gives the specifications of the V-100C Downconverter, which has an input impedance of 50 ohms and an output impedance of 75 ohms. The V-100R Video Receiver specifications are shown in Fig. 7-15. Further information may be had by contacting Vitalink Communications Corporation, 1330 Charleston Road, Mountain View, California, 94043.

DOWNLINK D-2X RECEIVER

The Downlink D-2X Receiver shown in Fig. 7-16 is manufactured as part of an overall TVRO Earth station system by Downlink, Inc. 30 Park

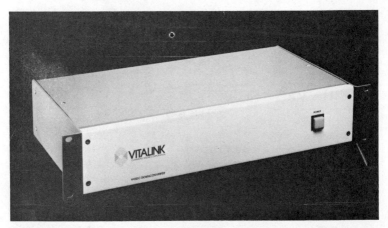

Fig. 7-13. The Vitalink V-100C block downconverter (courtesy Vitalink Communications Corp.).

140

```
V-100 C Down Converter

Input Frequency:    3.7 — 4.2 GHz        Output Frequency:   900 — 1400 MHz
Input Level:        −30 to −70 dBm       Output Level:       0 dBm
Input VSWR          1.5 : 1              Output VSWR         2 : 1
Input Connector:    Type N               Output Connector:   Type F
Input Impedance:    50 Ohms              Output Impedance:   75 Ohms
```

Fig. 7-14. Complete specifications for the V-100C downconverter (courtesy Vitalink Communications Corp.).

Street, Putnam, Connecticut, 06260. This receiver may be purchased individually in order to be used with a home constructed system.

The D-2X home satellite TV receiver features a compact, attractive control console which mounts at the television receiver location, along with a compact receiver which is mounted in a waterproof case and is designed for mounting at the antenna site. Figure 7-17 shows the master receiver section along with the remote console.

Normally, the receiver portion proper is located as close to the low noise amplifier as possible. Downlink's TVRO package uses a spherical antenna with the feed horn mounted on an aluminum mast some distance away. A mast adapter ring is mounted to the front of the receiver box as shown. This eliminates the need for expensive coaxial cable and improves television signal strength and picture quality. The internal circuitry of the receiver uses a phase-locked loop. This provides sharper picture images and brighter colors.

The D-2X allows multiple receivers to share one antenna/LNA combination, so viewers in different rooms of a house or in different apartments or condominiums can tune in different channels without the expense of separate antennas and a low noise amplifies. This is accomplished with dual dual conversion circuitry, which eliminates mutual interference between receivers.

As offered by Downlink, the D-2X receiver is also featured in the EP-2000 Electronics Package. This includes all components required for a complete home satellite television system, except for the antenna. Figure

```
V-100 R Video Receiver

Threshold:   8 dB C/N        Modulator:           CH 3 or 4
Channels:    24 Agile        Gain:                Automatic
Video:       1 v. P/P        Frequency Control:   Automatic
Subcarrier:  Any 4           De-Emphasis:         75 Microsec.
Audio:       600 Ohms        Clamping:            40 dB
```

Fig. 7-15. Complete specifications for the V-100 video receiver (courtesy Vitalink Communications Corp.).

Fig. 7-16. The Downlink D-2X receiver (courtesy Downlink, Inc.).

7-18 provides a complete table of specifications for the D-2X receiver. Figure 7-19 shows the remote control console located near the television receiver. Once the TV set is activated, all the viewer needs to do is select different satellite channels from the comfort of his easy chair.

MICRODYNE CORPORATION 1000 TVRN

Manufactured by Microdyne Corporation, P.O. Box 7213, Ocala, Florida, 32672, the 1000 TVRN is a satellite TV receiver and modulator combined in one unit. Shown in Fig. 7-20, the various TV channels are selected by rotating a manual digital readout switch assembly.

Fig. 7-17. Shown here is the receiver section with the remote console (courtesy Downlink, Inc.).

Input Frequency Range	3.7 - 4.2 GHz
Input Threshold	8.0 dB C/N
Downconverter	Dual Conversion, 70 MHz Output
Intermediate Frequency	70 MHz
Video Demodulation	Phase Locked Loop
Audio Demodulation	Phase Locked Loop
Video Tuning Range	3.7 - 4.2 GHz, Continuous
Audio Tuning Range	5.2 - 7.6 MHz, Continuous
Video Output Impedance	75 Ohms
Audio Output Impedance	Low
Video Output	NTSC Standard, 1V (P-P), Composite Video
Subcarrier Output	1V (P-P)
Power Requirements	120 VAC, 60 Hz
Power Consumption	25 Watts

Fig. 7-18. Specifications for the D-2X receiver (courtesy Downlink, Inc.).

Economy and convenience are provided by having the receiver and modulator circuitry located within the same case. The output from the 1000 TVRN may be fed directly to the television receiver. Both the receiver and modulator input/output may be accessed through the back panel, which is shown in Fig. 7-21.

The receiver features 24-channel selection, which is frequency synthesized. This means that no crystals are required for channel selection. The modulator is fitted with video and audio metering which is brought out to the front panel. Figure 7-22 gives the specifications for the single conversion, manually tuned receiver. Specifications on the modulator portion of the circuitry are shown in Fig. 7-23.

Fig. 7-19. Here, the remote console is located close to the television receiver (courtesy Downlink, Inc.).

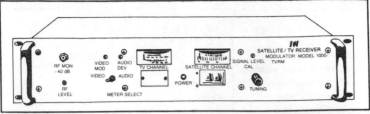

Fig. 7-20. The Microdyne 1000 TVRN receiver/modulator (courtesy Microdyne Corp.).

HR-100 SATELLITE RECEIVER

The HR-100 Satellite Television Receiver is manufactured by Satellite Supplies, Inc., 164B Gilman Avenue, Campbell, California 95008. Dubbed "The Entertainer", this receiver was designed by Taylor Howard, who is Adjunct Professor of Electrical Engineering at Stanford University. He is the acknowledged inventor of the first low-cost satellite receivers for personal Earth station use. He also chairs or participates as an active member on various NASA space probe teams and committees and was the recipient of the NASA Medal For Exceptional Scientific Achievement. The HR-100 satellite receiver is the culmination of years of pioneering design by Professor Howard to produce a satellite receiver that economically achieved extremely clear and dependable audio and video signals.

Shown in Fig. 7-24, the HR-100 offers 24-channel capacity, which allows full access to dozens of TV signals currently being transmitted by satellites. This offers the personal Earth station user a wide variety of new and unique programs which can be tuned in from around the world. The receiver is said to provide superior picture quality. Interferences that may affect terrestrial TV signals are often absent from satellite transmissions. Programs via satellite appear clear and crisp. The HR-100 uses advanced technology to provide high quality TV reception at an affordable price. The design has been field-tested under extreme conditions to insure maximum dependability and years of trouble-free operation.

The HR-100 is a compact receiver which is housed in an attractive mahogany cabinet which blends well with any leisure setting. Maximum TV

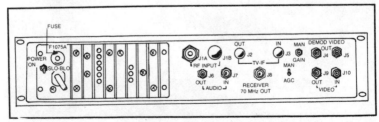

Fig. 7-21. Both the receiver and modulator input/output can be accessed through the back panel (courtesy Microdyne Corp.).

SYNTHESIZER:	
Stability	± 0.001% from 0° to 50°C; one part in 10^8 per three months
DOWN CONVERTER:	
Input Frequency	3.7—4.2 GHz
Input Impedance	50 ohms
RF Bandwidth	40 MHz nominal at 1 dB
Output Frequency	70 MHz
I-F DEMODULATOR:	
IF Frequency	70 MHz
IF Bandwidth	30 MHz
AGC Range	65 dB
Demodulator Type	FM
Demodulator Linearity	Linear to within ±1% over ±18 MHz
Video S/N vs C/N	Threshold occurs at<8.0 dB C/N ratio
VIDEO PERFORMANCE.	
Operating Parameters	Format: 525 / 60; System: M; fv Maximum: 4.25 MHz
Deviation Range	5 to 13 MHz peak at de-emphasis crossover frequency
Video Output Frequency Response	10 Hz to 4.25 MHz, ± 0.5 dB
Impedance	75 ohms
Level	1 volt peak-to-peak nominal
De-emphasis	525 lines per CCIR Rec. 405.1
Polarity	Black to white transitions positive going
Clamping	>40 dB for 30 Hz triangular dispersion waveform

NON-LINEAR DISTORTION:	
Differential Gain	±2% maximum, 10 to 90% APL
Differential Phase	Less than ±1°, 10 to 90% APL
2T Pulse Distortion	Less than 2%
LINEAR DISTORTION:	
Line Time Distortion	Less than ±1.5%
Field Time Distortion	Less than ±1.5%
AUDIO OUTPUTS:	
Subcarrier Frequency	6.8 MHz standard
Frequency Response	20 Hz to 15 KHz ± 0.5 dB
De-emphasis	75 usec time constant
Output Level	8 dBm nominal
OPTIONS:	
De-emphasis	PAL 625 lines and other de-emphasis filters available
Additional Audio Subcarriers	Up to two internal, however an external Model SCB-1 subcarrier demodulator is available to provide up to four separate audio subcarriers covering the range of from 4.5 to 7.5 MHz (See SCB-1 brochure)
Other I-F Bandwidths	17.5, 20, 25, 36 and 40 MHz in place of standard 30 MHz
Non-standard Audio Subcarrier	Available from 4.5 to 7.5 MHz in place of standard 6.8 MHz
Other TV Satellites	Available for 12 transponder operation in lieu of 24 channels

Fig. 7-22. Specifications for the receiver portion of this unit (courtesy Microdyne Corp.).

VIDEO:		Output	
Input Type	Composite video, sync. negative	Impedance	75 ohm unbalanced (VSWR 1.35:1)
Input Level	0.5v p-p minimum for 87.5% depth of modulation	Output Level	+40 dBmV to +60 dBmV continuously variable
Input Impedance	75 ohm, unbalanced	Spurious Outputs	60 dB below video carrier with video carrier at +60 dBmV and sound carrier at +45 dBmV
Frequency Response	±0.5 dB from 10 Hz to 4.2 MHz		
Differential Gain	±.2 dB maximum at 87.5% modulation	Sound Carrier Adjustment	Adjustable 10 to 20 dB below video carrier
Differential Phase	±0.5° maximum at 87.5% modulation	Frequency Tolerance Stability	±10 kHz VHF or mid-band
Hum and Noise	60 dB down with respect to 90% modulation	RF Monitor Level	−40 dB
Tilt	1% maximum on 60 Hz, square wave	Impedance	75 ohms
White Level Limit Adjustable	80% to 95% modulation depth	Vestigial Sideband Response (SAW filter)	−20 dB at channel edge −40 dB at adjacent picture and sound carrier frequencies and all frequencies farther removed from channel
AUDIO INPUT:			
Input Level	0 dBm (0.78 V rms) min to +10 dBm (2.50 V rms) max for 25 kHz deviation at 1 kHz rate		
		OPTIONAL I-F LOOP THRU:	
		*Input 75 ohms	+35 dBmV nom
Frequency Response	±0.5 dB from 30 Hz to 15 kHz	*Output 75 ohms	+35 dBmV nom
		* Jumpered via rear panel	
Carrier Shift with Modulation	±100 Hz or less	ENVIRONMENTAL:	
Harmonic Distortion	0.5% maximum, 30 Hz to 15 kHz at 25 kHz deviation	Temperature	Operating: 0° to 50°C Storage: −60° to 60°C
FM Hum and Noise	60 dB down with respect to 25 kHz deviation	Atmospheric Pressure	Operating: to 10,000 feet Storage: to 50,000 feet
Intercarrier Frequency Tolerance	Within ±500 Hz to being 4.5 MHz above video carrier (0° to +50°C)	Mechanical	3½"H × 19"W × 20"D
		Weight	Approximately 20 lbs.
		Prime Power	115V AC, 50 to 400 Hz
RF:		Power Consumption	Approximately 30 watts
Output Frequency	Any standard VHF channel; 2-13 and mid-band; A-1		

Fig. 7-23. Specifications for the modulator portion of the circuitry (courtesy Microdyne Corp.).

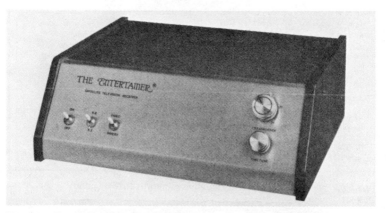

Fig. 7-24. The HR-100 satellite television receiver (courtesy Satellite Supplies, Inc.).

programming selection can be added to any permanent television location while providing a decorative addition to the viewing room.

One of the special features of the HR-100 satellite receiver is broadband automatic frequency control which locks the receiver frequency onto the desired channel. Also, a fine tuning control permits adjustment of the picture to peak conditions. A video invert switch allows viewing of programming which has been scrambled by inverted video transmissions. Dual downconverters are used to minimize interference from images and neighboring systems. The first downconverter is mounted at the antenna and reduces feed cable costs by 60%. A band-pass filter in the first downconverter removes image noise from the low noise amplifier output. Continuous standby power is provided which insures reliable operation in colder climates. Figure 7-25 gives the specifications of the HR-100.

AVANTEK AR1000 RECEIVER

The Avantek Model AR1000 Simulchannel satellite Earth station video receiving system is designed for performance, reliability, and economy in any application requiring broadcast quality, simultaneous reception of two or more program channels. This means that this receiver is intended for the commercial and CATV market.

Shown in Fig. 7-26, the complete system combines the Avantek ACA-4200 antenna-mounted LNA/downconverter which was discussed earlier with a flexible, modular i-f receiver/demodulator assembly. In an Earth station configuration using a dual-polarized antenna feed, two LNA/downconverters along with separate horizontal and vertical feed lines and a single AR1000 receiver will provide simultaneous reception of any six satellite-relayed video program channels with any combination of vertical and horizontal polarization. Four AR1000 receivers may be driven from the same feed line to provide total 24-channel simultaneous coverage.

147

Input Frequency	3.7 - 4.2 GHz	Dimensions	13½″=(345 mm) wide
Input Signal Level	−35 to −65 dBM		11″=(280 mm) deep
Noise Figure	13 dB (max)		5″=(125 mm) high
Output Video	1.0 Volt Peak to	Front Panel	■ Power on/off switch
	Peak (75 ohms)	Controls	■ Audio subcarrier 6.2 -
Output Audio	+ 10 dBM (600 ohm		6.8 MHz switch
	unbalanced)		■ Video invert switch
Audio Subcarriers	6.2 MHz & 6.8 MHz		■ Channel selector
	(others optional)		■ Fine tune control
Ac Power	115 Volts, 50/60 Hz,		
	20 watts		
First I-F Frequency	855 MHz		
Second I-F Frequency	70 MHz		

Fig. 7-25. Specifications for the HR-100 receiver (courtesy Satellite Supplies, Inc.).

The AR1000 receiver mainframe is compact and designed to be mounted in a standard 19-inch rack if desired. It includes one or two six-way power dividers and all backframe connectors and cabling to interconnect the system modules. The back of the unit is shown in Fig. 7-27. The power supply/digital control module accepts 115 Vac line power and provides all required operating voltages for the i-f receiver modules and LNA/downconverters. It also incorporates all required digital tuning logic as well as a front panel touch pad which is used to select and tune the i-f receiver modules. Each module is essentially a complete, digitally-tuned 940-1440 megahertz single-conversion video receiver with its own phase-locked loop demodulator. Connectors and terminals on the rear of the module provide access to the i-f input, video output, audio output, compositive baseband output, and agc level. The front panel of each module includes LED channel indicators, video lock indicator, and adjustments for video and audio output.

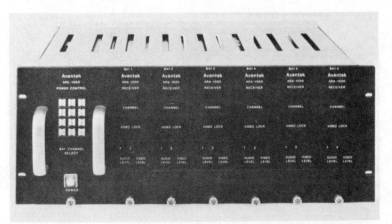

Fig. 7-26. The Avantek AR1000 simulchannel receiver (courtesy Avantek, Inc.).

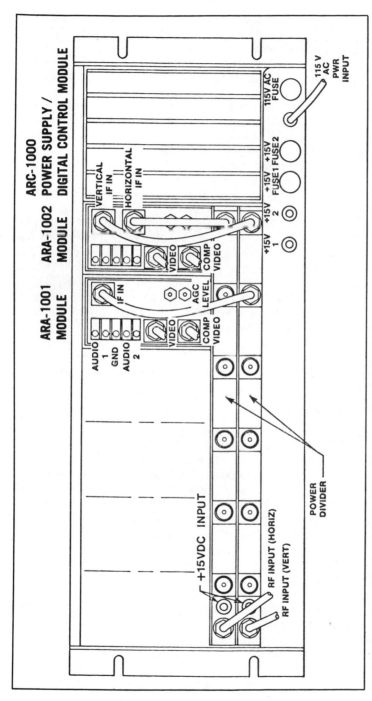

Fig. 7-27. Rear view of the Avantek AR1000 receiver (courtesy Avantek, Inc.).

149

The i-f receiver module is equipped with a single i-f input connector for use with single-feedline systems or for dual-feedline systems where it is desirable to shift the jumper cable from horizontal to vertical as required. The i-f receiver module is equipped with both a horizontal and vertical input connector and automatically selects the correct feedline for the channel to which it is tuned.

This model design accounts for the cost effectiveness of the AR1000 Simulchannel concept. A full-feature satellite video terminal requires only two LNA/downconverters, essentially the front-ends for all i-f receivers in the installation, and one power supply/digital control unit. The only components that are duplicated in the AR1000 are those that add to the performance and flexibility of the receiving system.

The i-f feedlines must be sufficiently long to run between the antenna and receiver installation should be ideally no more than approximately 10 dB loss at 1440 MHz and should be equipped with male type N connectors on both ends. Note that although the female type N connector on the ACA-4220 LNA /downconverter is weatherproof, the connector on the cable may not be weathertight and could allow moisture intrusion. Thus, it is a good practice to weatherproof the cable connections, possibly with a reenterable encapsulant or with surface-irradiated shrink tubing (the inner surface melts and flows around the connector when heated). Wrapping with plastic electrical tape does not provide a weatherproof seal.

If the coaxial feedline is to be buried at any point in the antenna-to-receiver run, make certain that the cable is of a type suitable for direct burial. Otherwise, it must be run in an underground conduit. Cable manufacturers can provide information on which of their products are suitable for direct burial.

Since the signals between the LNA/downconverter and the AR1000 receiver are carried in the 940-1440 MHz frequency range rather than at 3.7 to 4.2 GHz, there are a number of relatively inexpensive, readily available coaxial cables which may be used for the feedline runs.

The audio output of the i-f receiver modules is adjustable between −10 and +10 dBm, 600 ohms balanced, and is available at a barrier terminal strip on the rear panel of the modules. Connections to user-supplied amplifiers, switches or channel modulators may normally be made with virtually any twisted pair cables. Shielded twisted pair cables may be required to prevent noise or hum in particularly long runs or in locations with a particularly high radiated noise level. If the receiver modules are equipped with the optional dual audio subcarrier detectors, the second audio channel is also brought out to the barrier terminal strip.

Figure 7-28 shows a typical multiple access setup using the AR1000 receiver and the user-supplied antenna system. It can be seen that this is a complicated setup when compared to a system designed for personal TVRO Earth station uses. Figure 7-29 shows a multiple receiver design in block form. Again, this particular receiver and entire TVRO Earth station system is not economically applicable to the needs of a typical personal Earth

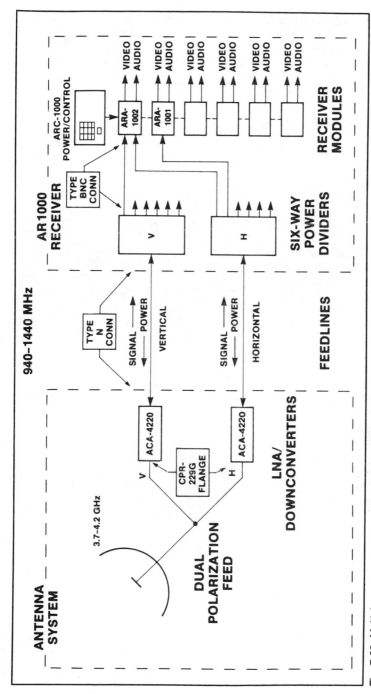

Fig. 7-28. Multiple access arrangement using the AR1000 receiver (courtesy Avantek, Inc.).

station enthusiast. The Avantek system is designed more for a commercial multiple-access application and for cable TV operations. This discussion is presented here to give the reader an idea of the equipment involved in commercial uses and for those individuals who may be contemplating the installation of a multiple access system for apartments, condominiums, etc. Figure 7-30 gives the specifications for the AR1000 receiver.

COMTECH MODEL 650 RECEIVER

Manufactured by Comtech Data Corporation, 613 South Rockford Drive, Tempe, Arizona, 85281, the Comtech Model 650 Receiver shown in Fig. 7-31 is antenna-mounted. Among the features included are digital channel select, a self-contained LNA supply, and a remote control option. The MGC/AGC switch permits the alignment of the antenna to insure optimum positioning. Included as a standard feature is a 6.2 or 6.8 MHz audio output channel, but optional audio demodulators permit different frequencies to be accessed.

If your installation utilizes a dual polarization antenna, the Comtech 650 will switch power between the vertically polarized channel and the horizontally polarized one automatically. A convenient disable permits the polarization to be locked in one position for use with a single polarization system or with stations which utilize the more common feed horn motorized rotator. Figure 7-32 gives the specifications for this satellite television receiver.

ICM TV-4000 SATELLITE RECEIVER

Manufactured by International Crystal Manufacturing Company, Inc., 10 North Lee, Oklahoma City, Oklahoma, 73102, the TV-4000 is available as a one-piece unit and with a remote control option. Shown in Fig. 7-33, this receiver is intended for the personal Earth station user and presently sells for less than $1,500. It offers 6.2 or 6.8 MHz audio output, which is switchable on the front panel. A relative signal level meter is mounted on the left side of the front panel and an accessory jack has been placed on the back panel for remote metering capability. This receiver uses Taylor Howard-designed demodulator circuitry and features a built-in automatic frequency control. There is a tunable audio subcarrier output jack for insertion of a tunable audio circuit option which sells for about $190. I-f bandwidth is 24 to 30 MHz and is set at the factory. A 15 Vdc power supply for connection to a low noise amplifier is available at an output jack on the back panel. The remove control option sells for approximately $110.

ICM TV-4400 SATELLITE RECEIVER

The same basic receiver as the previous one discussed, the TV-4400 satellite receiver provides a separate rf downconverter to be mounted at the antenna site. The entire system, then, consists of two discrete units. Shown in Fig. 7-34, the ICM system encloses the downconverter in an

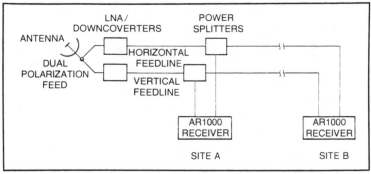

Fig. 7-29. Block diagram of a multiple receiver configuration (courtesy Avantek, Inc.).

environmentally protected box which is mounted at the antenna as close as possible to the LNA. It is designed to be mounted to the LNA bracket. The downconverter is a double-conversion device with input bandpass filtering.

The 3.7-4.2 GHz satellite signal is first converted to 850 MHz. It is then mixed with the local oscillator frequency of 920 MHz. The difference in these two frequencies is 70 MHz, which serves as the i-f output. The 70 MHz i-f is then fed from the antenna site to the i-f and baseband units by means of standard RG-59 coaxial cable or its equivalent. By converting from microwave to uhf frequency at the antenna site, cable losses at the high frequency are negligible. Tuning voltage is supplied through the RG-59 coaxial cable. Power is supplied by means of a separate two-conductor cable. A separate power supply for the downconverter may also be used. The baseband or tuning unit weighs 4½ pounds, while the downconverter weighs in at an even 1 pound. Price of the system as described is about $1,400 and an optional remote control console is available when the television receiver is located some distance from the Earth station. Figure 7-35 gives the specifications for the ICM TV-4400 satellite receiver. A separate modulator is required for connection to the television receiver.

VR-3X SATELLITE RECEIVER

Microwave Associates Communications' VR-3X Satellite Receiver is shown in Fig. 7-36. This is a low-cost commercial unit that offers high reliability and versatility combined with exceptional performance. This can be thought of as a single-channel receiver, although all 24 channels may be tuned by means of a screwdriver-adjust control. Afc (automatic frequency control) locks the receiver to the signal of the selected channel.

Standard features include a threshold extension demodulator that produces a 3 dB extension when compared to a 30 MHz i-f bandwidth. Baseband video or composite outputs can be provided thus allowing for additional flexibility.

RF Input[1]

Frequency range from block
downconverter ....940-1440 MHz
Tuning ....Digital tuning touch pad selects
I-F module, transponder channel
and feedline polarity[2]

Spacing between channels
(same polarization) ....40 MHz
Spacing between channels
(alternate polarization) ....20 MHz
Impedance ....50 ohms
VSWR ....1.5:1 maximum
Noise Figure ....15 dB maximum
Input Level /AGC range
(Receiver Module) ....−20 to −60 dBm
RF Connector (Power Divider) ....Type N

Video Output

Frequency ....10 Hz to 4.2 MHz ±0.5 dB
Impedance ....75 ohms unbalanced
Return Loss ....20 dB, minimum

154

Level	1.0 V p-p, ±0.5 V adjustment range
Polarity	Positive going, black-to-white
Deemphasis	525 line, CCIR recommendation
	405-1, (other standards optional)

Video Performance with threshold extension

Nonlinear distortion
Differential gain (10%-90% APL) .... ±2% max.
Differential phase (10%-90% APL) .... ±1°
2T Pulse Distortion 2%
Linear Distortion
Line time distortion .... ±1.5%
Field time distortion .... ±1.5%
Chrominance-to-luminance gain
inequality .... ±0.3 dB max.
12.5 T delay
inequality .... ±25 nanoseconds, max.

[1]For RF performance of associated LNA/downconverter refer to
ACA-4220 data sheet

[2]Automatic feedline selection only with ARA-1002 modules.

Fig. 7-30. Specifications for the AR1000 receiver (courtesy Avantek, Inc.).

Composite Baseband Output

Deemphasis ....525 Line, CCIR recommendation
405-1, (other standards optional)

Output Level ....6 mv/MHz Deviation
Impedance ....75Ω unbalanced
Return Loss ....20 dB minimum

Audio Output

Frequency ....20 Hz to 15 kHz, ±0.5 dB
Level (with standard test tone) ....−10 to +10 dBm,
adjustable range
Impedance ....600 ohms, balanced
Subcarrier frequency ....6.8 MHz standard
(others optional)
Harmonic distortion ....1%, maximum
Deemphasis ....75 microseconds
Connector ....Barrier strip

Threshold Characteristics (30 MHz Bandwidth)

Static (1 dB deviation
from straight line) ....<8 dB Carrier-to-noise, max.

Front Panel Controls

ARC-100 Control Module

Channel Display ....Two-digit, 7-segment LED
Module/Channel Select ....Touch pad (digits
0-9, Clear, and Enter)

I-F demod/processing module

Video Level ....Screwdriver adjust
Audio Level ....Screwdriver adjust
Video acquisition ....LED

Rear Panel Connections

ARA-1002 module

Video output ....Type BNC Female
Composite Baseband output ....Type BNC Female
Horizontal RF input ....Type BNC Female
Vertical RF input ....Type BNC Female
Audio output ....Barrier terminal strip

Environmental

Temperature (ARA-1001, ...0° to +50°C
ARA-1002, ARC-1000)   (+32° to +122°F)

Power Requirements ....115 VAC

Fig. 7-30. Specifications for the AR 10C0 receiver (courtesy Avantek, Inc.). (Continued from page 155.)

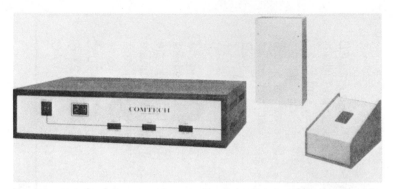

Fig. 7-31. The Comtech model 650 receiver (courtesy Comtech Data Corp.).

The VR-3X satellite receiver is designed for CATV and MATV (multiple access television) uses. It will feed a cable system directly when the optional cable modulator and frequency converter module are employed. The modules are simply plugged into the front of the unit after the panel is removed, as shown in Fig. 7-37. Figure 7-38 shows this receiver with its plug-in accessory board completely filled. The modulator and converter modules eliminate the need for an external cable modulator.

This receiver also offers the capability for i-f switching and provides a 4.5 MHz subcarrier output which consists of video plus program audio. This composite can be used to feed a cable system and a microwave system simultaneously. Figure 7-39 gives the specifications for the receiver and the optional cable modulator. Further information may be obtained by contacting Microwave Associates Communications, 121 Middlesex Turnpike, Burlington, Massachusetts, 01803.

VR-4XS SATELLITE RECEIVER

The VR-4XS satellite receiver shown in Fig. 7-40 is basically the same receiver just discussed but features a frequency agile synthesizer as the local oscillator and has front panel selectable 12 or 24-channel capability. Frequency selection is controlled either locally or externally, as determined by a front panel switch. A thumbwheel switch on the front panel supplies local control. External control is provided by means of binary-coded digital input to a rear panel connection. In the external control mode, the channel received is displayed on a front panel digital readout. A meter also located on the front panel monitors the automatic gain control voltage and is an indicator of relative received signal strength.

As with the previous receiver, a removable front panel allows easy access to the plug-in cards for maintenance, yet discourages unauthorized access when in place. Options include automatic vertical/horizontal polarization switching, either a video clamper or a video clamper/subcarrier processor card, single or dual audio demodulator cards, and a built-in cable

RF			**Controls**	
Frequency	3.7 - 4.2 GHz		Power On/Off	
Impedance	50 Ω		AGC/MGC Select	
Input Range	−60 to −30 dBm		Channel Select	
Nominal Return Loss	20 dB		Local/Remote Select	
Nominal Noise Figure	15 dB		Video Level Adjust	
Threshold	8 dB		Audio Level Adjust	
Connector	Type N		Manual Gain Adjust	
			6.2/6.8 MHz Select	
I-F				
Frequency	70 MHz		**Test Points**	
Bandwidth	30 MHz		AGC or MGC Signal Level	Terminal Strip
AGC Range	30 dB			
			Power	
Video			Prime	117 VAC ± 10%, 60 Hz Nom.
Format	525/50 or 60		Consumption	20 Watts
Frequency Response	± .5 dB, 15 Hz/4 MHz		LNA (unregulated)	18-24 VDC Terminal Strip
Polarity	Positive White			
Dispersion Removal	>40 dB		**Mechanical**	
Output	1 Volt P/P ± 3 dB		Control Box	4″ × 23″ × 15″ deep
Video Out	Type F		Size	
Composite Out	Type F		Antenna Mount	4″ × 16″ × 10″ deep
			Weight	18 pounds
Audio			Control Box	Table Top
Subcarrier	6.2/6.8 MHz standard		Mounting	
Frequency Response	± .5 dB, 15 Hz to 15 KHz		Antenna Mount	Antenna Frame
De-emphasis	75 micro seconds			
Impedance	600 Ω balanced		**Options**	
Level	0 dBm ± 3 dB		Remote Control	
Harmonic distortion	< 1%		5.8 /7.4 /7.6 MHz program demods	
Connector	Terminal Strip		Modulator, channel 3 or 4 selectable	

Fig. 7-32. Specifications for the model 650 receiver (courtesy Comtech Data Corp.).

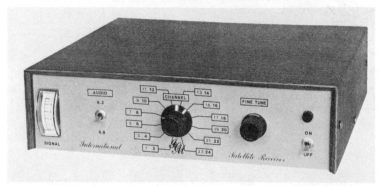

Fig. 7-33. The ICM TV-4000 satellite receiver (courtesy International Crystal Mfg. Co.).

modulator consisting of an i-f modulator and an rf converter card. The specifications given in Fig. 7-39 generally apply to the VR-4XS satellite receiver.

BRIEF NOTES ON OTHER RECEIVERS

There are many other receivers on the market today. Some have only been recently offered, while others will come on the market in the very near future. One of these is the Gillaspie & Associates Model 8000 shown in Fig. 7-41. From the information I have been able to obtain on this design, it would seem to be a beautiful little unit that can accurately be described as a "top-of-the-line" receiver for personal Earth station use. Fully synthesized, this microprocessor-controlled receiver provides pushbutton channel selection, wireless remote control, full-range audio, video and audio scan, and is tastefully designed with a smoked glass front. Further information should be available by the time this book is published, so a letter to Gillaspie & Associates, Inc. (address provided elsewhere in this chapter) should get you a complete information package.

Fig. 7-34. The ICM TV-4400 Satellite receiver (courtesy International Crystal Mfg. Co.).

				Baseband	Downconverter
Frequency	3.7 to 4.2 GHz				
Noise Figure	12 dB typical	Size:		3½ × 8 × 8	3 × 4 × 7
Input Impedance	50 ohms nominal	Weight:		4.5 #	1 #
Intermediate Freq.	850 and 70 mHz (dual conversion)	Input Jack	Type N		
I-F Bandwidth	25 mHz	Input Level	Optimum −25 to −40		
Video Output	1 volt peak to peak, 75 ohms	Image Rejection	See Note below		
Frequency Response	10 Hz to 4.2 MHz plus or minus 0.5 dB				
De Emphasis	525 Line per CCIR recommendation 405-1	Meter Output	0 to 1 milliampere. −50 dBm approx ¾ scale		
Audio Outputs	0 dBm, 600 ohms internally settable	Output Jacks	Standard RCA phono plugs		
Frequencies	6.2 and 6.8 MHz subcarriers standard				
De Emphasis	75 microseconds	Remote Control	Optional Remote control — receiver equipped with standard jack.		
Operating Temp.	0 to 70 degrees C				
Power	115 volts ac 50 to 60 Hz, 15 watts nominal				

Fig. 7-35. Specifications for the ICM TV-4400 receiver (courtesy International Crystal Mfg. Co.).

Recently offered by the Skyscan Corporation, 250 East 36th Street, Tucson, Arizona, 85713, is the System SS-6900, which is a complete Earth station that incorporates a pushbutton receiver, as shown in Fig. 7-42, and a wireless remote control. No hard wiring is connected between the receiver and the control unit, the latter being a hand-held design very similar to those used for the remote channel switching of standard television receivers. This is a relatively inexpensive system which includes a parabolic dish antenna with a steel base. Total price is about $4,000 with everything needed for installation, except the concrete base which is user-supplied.

Hustler, Inc. has recently introduced a 24-channel satellite receiver named the Model SVS-1000. Designed for the Earth station component market, this receiver is designed to pick up transmissions in the 3.7 to 4.2 GHz range with an audio subcarrier of either 6.2 or 6.8 MHz. The various controls include subcarrier frequency, fine tuning, audio level, and a 24-position channel selector. The receiver includes a built-in regulated power

Fig. 7-36. The VR-3X satellite receiver (courtesy Microwave Associates Communications).

Fig. 7-37. Modules may be plugged into the VR-3X receiver by removing the front panel (courtesy Microwave Associates Communications).

supply for the LNA and its own self-contained, switchable modulator with a 75 ohm output on vhf channel 3 or 4. This output is used to feed the signal to a standard television receiver's antenna terminals.

KLM's Sky Eye II satellite receiver is shown in Fig. 7-43. It is a two-part system which is easy to install and use. The compact control console shown in the foreground is installed at the television set location. The modular receiver unit mounts at or in close proximity to the TVRO antenna.

The Sky Eye II kit comes in simple, easy to assemble sections. All microwave circuitry is factory-wired and tested. The assembly manual is clearly written. Specifications include single conversion/image rejection circuits (to easily handle strong and weak signals without instability or distortion); built-in dc block for feedline powered LNA; full video tuning 3.7 to 4.2 GHz; audio tuning at 5.5 to 7.5 MHz; separate regulated power supplies for the low noise amplifier and the receiver; optional polarity control; and control console to receiver cabling.

The KLM Sky Eye II receiver is available both as a kit and in a factory assembled version. The kit sells for about $700, while the wired model will cost about $1,400. Further information may be obtained by writing KLM Electronics, Inc., P.O. Box 816, Morgan Hill, California, 95037. Another chapter in this book will provide greater detail on the assembly of the KLM Sky Eye II kit.

Another satellite receiver which is popular among TVRO Earth station enthusiasts is the Avcom Com3 shown in Fig. 7-44. It offers scanning of all 24 satellite channels in the same manner that a police band receiver scans its channels. Switch-selectable tuning, automatic frequency control, and remote control capability are also offered. This is a high-quality receiver which offers attractive styling and an excellent receive threshold for good picture reception. This receiver is quite a bit more expensive than others that are designed for personal Earth station use. List price is about $3,000, although I have seen them advertised in discount catalogs for about $500 less. Additional information may be obtained by writing Long's Electronics, P.O Box 11347, Birmingham, Alabama, 35202. This company is a distributor for Avcom products.

ACCESSORIES

As is the case with any popular field, there are many options and attachments that may be purchased for your TVRO Earth station. One of these is the ICM Signal Purifier, which is manufactured by International Crystal Manufacturing Company, Inc. Shown in Fig. 7-45, this circuit sets the signal level to the satellite receiver with gain which is adjustable from +5 to −10 dB. A five-pole bandpass filter eliminates interference from out-of-band transmitters and radar systems. The purifier is said to improve picture quality on all inexpensive receiver/LNA combinations by filtering out image noise. The more expensive receivers may already incorporate this filtering circuit.

TVRO Earth station installations generally have two problems remaining once a picture is finally obtained. These problems can be defined by answering the following questions:

■ Can the picture be improved by more or less signal?
■ Is out of band or image interference degrading the picture?

The first question identifies a two-part problem. The first is weak signal at the receiver, and the second part is signal overload at the receiver's front end, which causes distortion. The second question alludes to interference from other services such as radio transmitters and radar installations.

These problems are at their peak in installations where cable runs are either very long or very short and where inexpensive receivers are in use. The ICM Purifier helps solve these difficulties. Basically, it is a five-pole bandpass filter (3 dB-bandwidth at 3500 to 4500 MHz) followed by an amplifier. The noise factor of this device is better than any passive mixer and the gain is adjustable.

The purifier has a male (N) connector on the output end, and a female connector is used at the input. Power is supplied through a two-wire line

Fig. 7-38. Here, the VR-3X's plug-in accessory board is filled completely (courtesy Microwave Associates Communications).

163

GENERAL

Type	superheterodyne, dual conversion
Radio Capacity	525 line video plus audio subcarriers
Frequency Range	3.7 to 4.2 GHz*
Noise Figure	15 dB nominal
Threshold	3 dB extension when compared to 30 MHz BW—i.e., 7 dB threshold
Local Oscillator Stability	w/AFC
Image Rejection	greater than 40 dB
RF Return Loss	20 dB
RF Input Level	−30 dBm maximum
I-F Bandwidth (70 MHz)	30 MHz standard
Modulation	FM
Video	
Outputs	one baseband output, two filtered video outputs or composite** outputs
Level	1 V P-P, adjustable to 1.5V P-P***
Impedance	75 ohms
Return Loss	26 dB
Audio	
Output	+9 dBm maximum test tone
Impedance	600 ohms balanced
Frequency Response	
30 Hz to 15 kHz	±0.5 dB

VIDEO PERFORMANCE
(w/Threshold Extension)

Signal-to-Noise (CCIR Weighted at 16 dB C/N, 10.75 MHz peak deviation)	53 dB minimum;
Signal-to-Hum (P-P to RMS)	58 dB
Clamping (Dispersion Waveform)	50 dB minimum rejection
Differential Gain (10-90% APL)	±2%
Differential Phase (10-90% APL)	±1°
Line Time Distortion	1 IRE unit maximum
Field Square Distortion	2 IRE units maximum
Short Time Distortion	3 IRE units maximum
Frequency Response	
Video (Filtered) Output	
10 kHz to 4.2 MHz	+0.5, −3 dB maximum
Baseband Output	
10 kHz to 4.2 MHz	+0.5, −1 dB maximum
4.2 MHz to 8 MHz	+0.5, −3 dB maximum

PRIMARY POWER

Source	105 to 125 Vac (47 to 63 Hz) (210 to 250 Vac Optional)
Power Consumption	
Basic Configurations	less than 45W
All Options	less than 75W

*Local Oscillator field adjustable through front panel.
**Composite output is filtered video with 4.5 MHz aural subcarrier.

Fig. 7-39. Specifications for the VR-3X receiver and optional cable modulator (courtesy Microwave Associates Communications).

ENVIRONMENTAL

Ambient Temperature

Full Specifications	+ 10 to +40°C
Operational	0 to +50°C
Humidity	90% (0 to +50°C)

Elevation

Operational	15,000 ft / 4,500m
Storage	50,000 ft / 15,000m

MECHANICAL

Size	19" (w) × 18" (d) × 5.25" (h)
	(48.3 × 45.7 × 13.3 cm)
Weight	less than 25 lbs. (11.4 kg)

Connectors

Audio	terminal strip
RF Input	Type N Coax
Cable Modulator Output	Type F Female
Video Baseband and	
Filtered Outputs	Type BNC Coax
Primo Power	AC Power Cord w/3-prong plug
Auxiliary Power Output	4 pin Molex Female,
	20 to 26 (positive unregulated) Vdc
	200 mA

COMPOSITE OUTPUT
(Meets Video Performance Specified Except As Follows)

Video

Short Time Distortion	5 IRE ringing
Frequency Response	4.5 MHz Notch, −3 dB
	at 4.2 MHz
Audio	no measurable degradation

CABLE MODULATOR (Optional)

Modulation Flatness

30 Hz to 4 MHz	±0.5 dB
4 MHz to 4.2 MHz	+1 dB, −2 dB
Differential Phase (10-90%)****	±1°
Differential Gain (10-90%)****	0.5 dB maximum
AC Hum and Noise****	60 dB down
Lower Sideband Vestigial	meets FCC 73.687
Frequency Stability	±0.005% maximum

RF Output

Level	+55 dBmV typical;
	+50 dBmV minimum
Impedance	75 ohms unbalanced
Return Loss	16 dB minimum
Output Frequency	any standard CCIR
	system "M" channel, T7 through T10
	and 2 through W
Spurious Signals and	
Intermodulation Products	at least 60 dB below
	visual carrier

***Not adjustable with Cable Modulator.
****87.5% modulation.
All specifications are subject to change without notice.

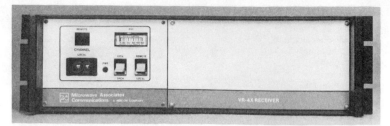

Fig. 7-40. The VR-4Xs satellite receiver (courtesy Microwave Associates Communications).

and can be anywhere between 10 and 24 Vdc at slightly less than 20 milliamperes. Gain is adjustable with a small screwdriver and is set for optimum system performance. This adjustment is shown in Fig. 7-46.

Installation is very simple and its explanation will be better understood by referring to Fig. 7-47. The purifier should be placed in the antenna feedline at the receiver. If the dc isolation function is to be used, the yellow/white wire must be attached to the +LNA/power supply connection to provide power for LNA. The purifier is designed for indoor use only. Connect the red wire to a source of +dc voltage in the 12 to 24 volt range. Your present LNA power supply will be satisfactory. Current drain is about 20 MA at maximum.

The gain potentiometer under the snap on the cover is adjusted with a small screwdriver for best picture quality with your system. This adjustment may vary slightly with different transponders and satellites. The amount of improvement obtained with the installation of this device will vary depending upon each individual system. The purifier will not solve problems which are caused by marginal signal strength in band interference or detectors.

Gillaspie Ampli-Splitter

Sometimes, two neighbors may wish to go together to purchase a satellite Earth station. If more than one satellite channel is to be received, the fact that two persons are using the same station presents several problems, the main one of which is the fact that there is a single

Fig. 7-41. The Gillaspie Model 8000 (courtesy Gillaspie & Associates).

Fig. 7-42. The Skyscan pushbutton receiver and wireless remote control unit (courtesy Skyscan Corp.).

receiver. Whatever station one neighbor is watching must also be viewed by the other. The Gillaspie Model ASP-100 Ampli-Splitter can help to overcome this problem. Shown in Fig. 7-48, this device is designed to be

Fig. 7-43. The Sky Eye I satellite receiver (courtesy KLM Electronics, Inc.).

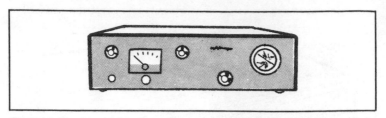

Fig. 7-44. The Avcom Com 3 satellite receiver (courtesy Long's Electronics).

installed into a 3.7 to 4.2 GHz satellite television system when two or more receivers are to be connected to the same antenna/LNA. This setup is shown in Fig. 7-49A. The Ampli-Splitter allows one antenna/LNA to provide output to two separate receiving stations. Since the antenna is often the major portion of a TVRO system when referring to capital outlay, a great deal of savings can be had here. It is necessary to purchase a separate satellite receiver for each user as well as a separate downconverter. The Ampli-Splitter contains an amplifier and a special two-way signal splitter that has a minimum isolation of 50 dB between its two output ports. This prevents the local oscillators of adjacent receivers from interfering with each other. The Ampli-Splitter can be powered either from the receivers or from another source through a built-in dc block and feed-through capacitor. The amplifier section has a gain of at least 6 dB and can be used to stretch the antenna receiver distance.

Only one Ampli-Splitter is required for a two-station setup. In single quantities, it sells for about $500. Figure 7-49B gives the preliminary specifications for this device.

ICM Tunable Audio

The ICM Tunable Audio device is housed in a small aluminum box and is designed for use with all satellite receivers having a subcarrier or unfiltered viewo output. Made by International Crystal Manufacturing Company, this device is shown in Fig. 7-50. BNC and phono inputs are supplied at the back panel. Input levels in the −30 dB to +10 dB range are

Fig. 7-45. The ICM signal purifier (courtesy International Crystal Mfg. Co.).

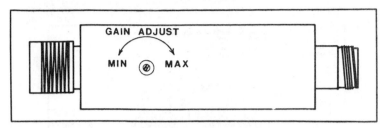

Fig. 7-46. This gain adjustment on the ICM signal purifier allows for optimum system performance (courtesy International Crystal Mfg. Co.).

acceptable. This unit allows for the tuning of audio channels which are contained on the subcarriers of satellite transmissions. Single-knob tuning with a stereo indicator light makes operation extremely simple. Both mono and stereo outputs are provided at a 1 V peak-to-peak nominal level. Output conectors are RCA phono type. Unit price of the ICM Tunable Audio device is about $190.

Avantek Line Extender

A line extender is a device which boosts the signal strength of Earth terminal installations which have high loss between the antenna and the first down conversion circuit. The Ampli-Splitter formerly discussed was a combination splitter and line amplifier circuit. It offers only marginal line gain, however. Its main purpose is to split the output signal of the antenna/LNA. The Avantek Line Extender (Model ALD 4200, 4201) boosts the signal by a larger amount. Signal loss can result from long cable runs or extensive power splitting in TVRO systems. The Line Extender is available in two power configurations for adaptation to dc cable-powered or integral ac-powered LNA installations. Figure 7-51 shows the connection of the Line Extender in a typical TVRO Earth station. Notice that it is inserted between the LNA and the power block. The increased signal strength provided by this amplifier circuit allows for the final output signal to be split many times. Figure 7-52 gives the specifications for this miniature device, which is housed in a weatherproof aluminum container. Figure 7-53 shows the basic unit along with its dimensions.

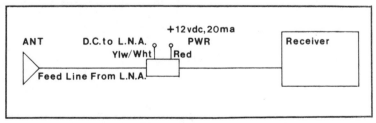

Fig. 7-47. The purifier is installed in the antenna feedline at the receiver, as shown here (courtesy International Crystal Mfg. Co.).

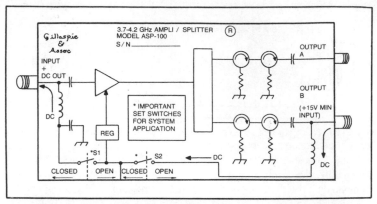

Fig. 7-48. The Gillaspie Model ASP-100 ampli-splitter (courtesy Gillaspie & Associates).

There are many other accessories available for TVRO Earth station users. Many of these are presently being offered by various discount mail order houses. From these companies, you will find a wide selection of TVRO equipment, such as antennas, receivers, LNAs, and the various hardware associated with these systems. Listed in Long's Electronics latest catalog is a Javelin JRF-RV Modulator which is designed to accept the output (both audio and video) from a satellite receiver. Upon entering the unit, the internal circuitry converts the signals so that they can be used with any standard television receiver. This particular unit may be tuned to any

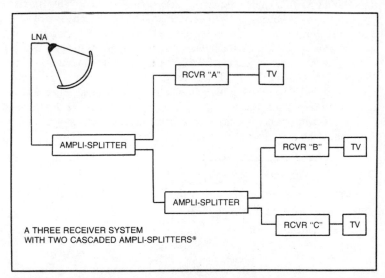

Fig. 7-49A. Using the ampli-splitter, one antenna/LNA provides output to two receiving stations (courtesy Gillaspie & Associates).

PRELIMINARY SPECIFICATIONS (3.7-4.2 GHz, 25 C AMBIENT)
Gain ....6dB Min.
Noise Figure ....7 dB Typical
Isolation ....50dB Min.
Power Requirements ....+15v@50mA + LNA power.
Built in dc block so power can be fed from the
receiver to the AMPLI-SPLITTER and LNA.
Input-Output connectors: Type-N Female

Fig. 7-49B. Preliminary specifications for the ampli-splitter (courtesy Gillaspie & Associates).

vhf television channel between 2 and 6 and operates on 110 Vac house current. List price for the Javelin Modulator is only $140.

This same catalog also offers a separate feed horn which might be purchased by those individuals who would like to build their own satellite TV antenna. Shown in Fig. 7-54, the feed horn pre-selects and directs signals to the low noise amplifier and is said to produce excellent audio and video reception. It is designed for mounting to the LNA bracket. Of course, coaxial cable will be needed for interconnecting the various pieces of equipment. RG-217 cable is often used for TVRO purposes. It lists in most

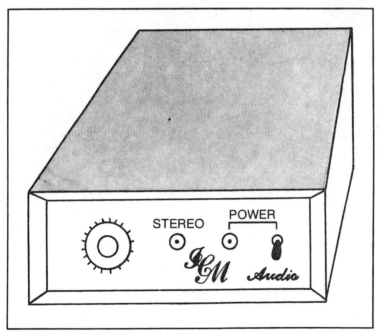

Fig. 7-50. The ICM tunable audio (courtesy International Crystal Mfg. Co.).

171

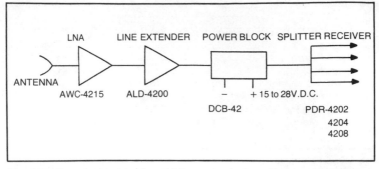

Fig. 7-51. Connection of the Avantek line extender in an Earth station (courtesy Avantek, Inc.).

catalogs for about $1.40 per foot and can often be purchased in many different lengths and fitted with the proper connectors to mate with your specific equipment.

As the interest in TVRO Earth stations continues to grow, more and more companies will be offering system components along with the many options. Right now, you pretty much have to know where to look as well as where to write to get the specific information needed on products which fall into the TVRO line.

SUMMARY

Even though the use of TVRO Earth station equipment is fairly small as of this writing, especially when compared to all of the owners of television sets in the United States alone, there are many products which should be reviewed by the TVRO enthusiast before a purchase is made. This chapter has overviewed a sampling of these products and provided specifications wherever possible. As newer equipment is offered, a comparison may be made in system specifications and the reader should be able to determine the quality of units not specifically dealt with in these pages.

I have purposely covered several pieces of commercial-grade equipment in order to show a specification comparison with those devices which are intended strictly for the personal Earth station market. Also, some readers may wish to install their own multiple access system in apartment

Model	Freq. (GHz)	Gain (dB min.)	Gain Flatness (dB max)	Noise Figure (dB max)	Power Output (dBm min)	Intercept Point (dBm typ.)	VSWR (in/out)	Power
								+15 to +28
ALD-4200	3.7-4.2	20	± 0.5	8.0	+3	+13	1.5	VDC 70 ma
ALD-4201	3.7-4.2	20	± 0.5	8.0	+3	+13	1.5	115 VAC

* D.C. power fed on cable. Power insertion block. DCB-42, may be required where used with receivers which do not provide dc power on the cable.

Fig. 7-52. Specifications for the line extender (courtesy Avantek, Inc.).

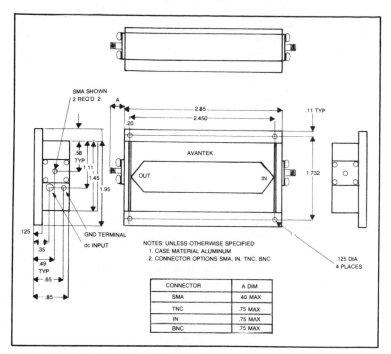

Fig. 7-53. Block diagram and dimensions of the line extender (courtesy Avantek, Inc.).

complexes. It should be remembered that as the state of the art continues, a lot of this commercial equipment will eventually be replaced by the CATV companies. It may then be sold through surplus outlets and can be an excellent buy for personal Earth station builders and owners.

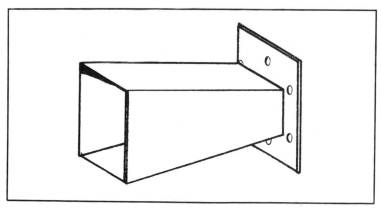

Fig. 7-54. A separate feed horn is available for those who wish to build their own antenna (courtesy Long's Electronics).

I have not meant to recommend one manufacturer's products over another. Some equipment is superior when compared to other makes, but it also costs more for these advantages. Personal needs and bank accounts will dictate the final selection, which should be made after a market-wide comparison of products has been completed and the manufacturers consulted for further detailed information.

Chapter 8
TVRO Systems

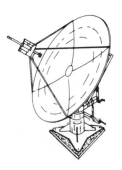

The previous chapter dealt with individual components found in TVRO systems. It was mentioned several times that many of these components were also offered in a complete systems package which included everything to assemble a fully operational personal TVRO Earth station. Many companies are moving in this direction with an emphasis upon providing everything that is needed for a complete installation. There are many advantages to this type of package.

When Earth stations are assembled from discrete components made by many different manufacturers, problems sometimes develop. These can include input and output connectors which do not mate with the rest of the system, low noise amplifiers which may not provide adequate noise ratings for the antennas used, receiver output which is not adequate to a certain type of modulator, and so on ad infinitum. These problems incurred by the Earth station builder/owner are often passed on the various manufacturers when questions begin to pour in and refunds are asked for due to equipment incompatibility.

When the manufacturer supplies the complete package, which basically consists of antenna/mount, feed horn/LNA, receiver/modulator, remote control, and all interconnecting cables, the customer as well as the supplier can be assured that a completely compatible system of components is being received. Any interconnection problems are worked out in advance. The customer feels a little more secure in knowing that he is purchasing a *complete* TVRO package with an assembly manual. It takes a lot of the guesswork out of the entire installation process.

But can most manufacturers make all of the equipment themselves? The answer is usually no. In talking with several companies which offer TVRO packages, I found that most of them manufacture a receiver, modulator, or downconverter and purchase the remainder of the components

(LNA, antenna, line amplifier) from companies which concentrate on the manufacture of these items alone. Often, the company which offers the package will place its own label or trademark on the components which have been purchased from other manufacturers. This way of doing business can have cost advantages for both the package supplier and the customer. The supplier will usually not buy in single-unit quantities. He will buy many companies and will receive a quantity discount. Some of this discount is often passed along to the purchaser of a complete Earth station package. A good example of this is seen in a pricing list from Gillaspie & Associates, Inc. in Monterey, California. A receiver which is priced at about $1,300 sells for only $1,000 when purchased in quantities of ten or more. Additionally, they offer a 120° low noise amplifier for $700 when purchased alone, but if you buy one of their receivers, you can get the LNA for only $650, which is the dealer's price. Gillaspie & Associates does not make a commercial low noise amplifier, so all they're interested in really is selling their receivers. If they can offer an LNA at close to cost, they will probably sell more receivers. This is financially attractive to them as well as to their customers. This company also offers a complete TVRO package consisting of the receiver, downconverter, 12-foot parabolic antenna and feed horn, low noise amplifier, modulator, and 75 feet of coaxial cable along with the power supply for a package price. The parabolic antenna may be omitted, which will save the purchaser $2,500, but of course you will have to buy another one elsewhere.

While researching this book, the author talked with a dealer representative from Channel Master who has recently offered a complete TVRO Earth Station package. This excellent station was being demonstrated at the time of our conversation. The representative stated that the components were manufactured by other companies but were being offered as a complete installation by his company which was fast in the process of gearing up to manufacture their own system. Of course, Channel Master provides the warranty and assumes service responsibility for all of the equipment it sells. Certainly, anyone could purchase these components separately, but assembly would probably not be as simple, since many of them were not *specifically* designed for easy interconnection.

Channel Master has long been known in the electronics field for offering excellent consumer products. They are probably a much larger company than the manufacturers they get their components from. Having a major corporation behind you helps considerably. Channel Master maintains a local inventory and service department, which can save the TVRO Earth station owner a lot of long distance phone calls which are necessary when dealing with factory suppliers and distribution warehouses. Due to the specifically constructed Earth station package, all components are tested and matched for system installation. This provides a complete turn-key station that arrives together and works together.

Channel Master usually sells to local distributors such as electronics stores. Chances are, the purchaser of this equipment would be dealing with them. Channel Master supplies factory training for all distributors, and company technicians are available to assist them in marketing, installation, and operation of Channel Master satellite reception equipment. Distributor-sponsored training sessions are designed to acquaint all interested personnel, such as the potential Earth station buyer, with satellite technology, equipment, and installation techniques. The distributor you buy your equipment from is offered computer assistance for calculating site coordinates and also for obtaining information on any potential microwave interference at a given site. All of this amounts to a lot more service with the sale than can be fairly expected from the manufacturer of one or two Earth station components. The latter would probably prefer to sell to a company like Channel Master rather than to deal directly with the public. More units can be sold outright and everyone can be assured that the customers' questions and needs are adequately answered.

Another method of buying TVRO packages is to consult catalogs of mail order companies which sell electronic communications equipment. Amateur radio outlets have recently turned to offering complete Earth stations in addition to discrete components. Long's Electronics in Birmingham, Alabama has two packaged systems. System I sells for about $3,100 and includes a ten-foot fiberglass dish antenna with polar mount and an Avantek 120° low noise amplifier, the KLM Sky Eye I satellite receiver, antenna feed horn, LNA mount, and Javelin rf modulator. The customer must supply the RG-59 cable. The package is also available without the modulator for about $3,000. Total list price of all system components would be in excess of $4,000 if purchased individually. It should be noted here that the prices quoted were the ones in effect for the latter part of 1981 and may reflect a special sale by the supplier.

System II offered by Long's Electronics is listed at $5,995 and includes the Avcom Com3 receiver, Avantek LNA, rf modulator, coaxial cable, feed horn, LNA mount, and an 11-foot aluminum dish with polar mount. List price of these components when purchased separately would be nearly $7,500. This reflects a savings of nearly $1,500. Again, lower prices can often be offered by large volume mail order outfits because they buy in quantity.

Heathkit is probably the best known supplier of electronics components designed for consumer use in the world today. This is especially true when speaking of amateur radio operators, home computer enthusiasts, and home experimenters in general who like to assemble their own equipment. In their Fall 1981 catalog, Heathkit has announced their entry into the TVRO Earth station market by offering a complete home Earth station kit. Scientific-Atlanta is supplying many of the components, all of which are warranted and serviced by the Heath Company.

As usual, Heathkit takes their buyers through every stage of developing a complete finished product. The first step is their site survey kit which sells for $30. This will help the potential buyer determine whether a Heathkit Earth station can be installed at his site. When you order this kit, you will receive a computer printout which shows satellite "look angles" for your address, a compass for determining the directions of different satellites, and an inclinometer to help determine the degree of elevation between the Earth's surface and the satellite. All of this is done to assure a clear line-of-sight path to the satellite.

You also get a manual which explains satellite TV and the Heathkit Earth station. It further includes information on how to use the compass and inclinometer, how to find the right ground conditions, and how to properly configure your Earth station. A copy of "Sat Guide" is also included to show the vast selection of programming available from just one satellite (SATCOM I). If you decide to order the Earth station package, the $30 site survey kit price is refunded.

Heathkit is especially noted for its concise assembly manuals. These provide easy-to-follow instructions and are backed up by technical assistance which is only a letter or telephone call away. Heathkit states that the integrated low noise amplifier/downconverter, 3-meter satellite antenna, and receiver electronics are made by Scientific Atlanta. The satellite receiver is equipped with its own modulator, so its output may be connected directly to your television set. Features include drift-free reception, digital channel selection, memory, twelve-hour digital clock, and built-in security circuit to which your own alarm may be attached. A wireless remote control unit is included in this kit. It is built by Zenith and puts you in complete control of receiver functions from your arm chair.

The construction portion of the kit includes assembly of the 3-meter antenna, and an earth foundation kit enables you to anchor this antenna in the ground firmly enough to withstand winds as high as 100 mph. This is a tripod arrangement which consists of digging three holes and filling them with concrete. You also build the rotatable feed assembly, which channels the signal to the low noise amplifier and downconverter. It consists of a rotor which automatically adjusts the feed horn to the proper polarization for each channel. The LNA/downconverter is preassembled and comes as a finished unit.

The Heathkit satellite TV receiver kit features electronic tuning of 24 channels, and a special built-in memory allows instant switching between any two preselected channels. The receiver can be programmed to turn itself on at times the owner selects for viewing or recording purposes. It is housed in an attractive cabinet with walnut panels and has separate connections for a television set and/or video cassette recorder. The built-in security circuit allows for connection to an alarm (customer-supplied) which will sound when an attempt is made to remove the LNA/downconverter.

Downlink, Inc. is one of the leading companies in the TVRO field and they manufacture most of their own equipment, with possibly the exception of the low noise amplifier. Downlink is offering Skyview I and Skyview II kits for $5,000 and $7,500 respectively (installed) in knockdown kits for relatively easy assembly. Again, Downlink does most of the manufacturing

Fig. 8-1. Downlink President Portus Barlow (Right) talking with Robert Cooper (Left) at a satellite seminar held in 1981 in Washington, D.C. (courtesy Downlink, Inc.).

itself. They even make the wooden enclosures for their receivers. While researching this book, I talked often with Downlink President, Portus Barlow, shown in Fig. 8-1 (on the right) talking with Robert Cooper at the Downlink booth which was set up for the Satellite Private Terminal Seminar in Washington, D.C. Figure 8-2 shows the Downlink antenna prototyping shop, where a variety of antennas are built and tested, such as the one-piece weld feed horn under construction. Their 13,000 square foot antenna manufacturing facility is presently turning out over 500 low-cost spherical antenna kits per month, as shown in Fig. 8-3.

While most of Downlink's discrete components are discussed elsewhere in this book, a short system overview will be presented here. The Skyview I system comes as either an 8 or 12-foot spherical antenna made of aluminum screen reflector surface and a supporting frame of redwood strips, angle irons, and galvanized tuning bolts. Up to seven of the currently active satellites may be simultaneously received by the spherical antenna without actually moving the dish. Instead, the remote mounted feed horn/LNA assembly is moved.

The Skyview II kit is mass-produced from 4×4 lightweight modular panels that form 8, 12, or 16-foot spherical antennas. All of these have plastic aluminized surfaces and are injection molded. Included in the systems kit is an all weatherproof downconverter/receiver with an enclosure that contains the receiver electronics. A remote control console is included which is mounted on the television set, while the downconverter receiver is mounted at the antenna. This system is shown in Fig. 8-4. There is also a new modular receiver available called the D3. This is connected to a

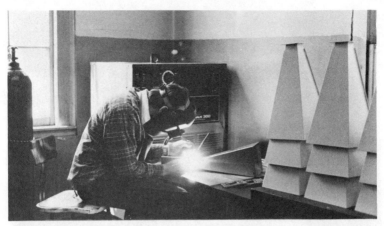

Fig. 8-2. Downlink's prototyping shop, where a variety of antennas are built. Shown here under construction is a one-piece weld feed horn (courtesy Downlink, Inc.).

conventional low noise amplifier or the new Downlink integrated downconverter that drops the 4 GHz signals to 70 MHz, eliminating the normal signal loss through coaxial cables and couplers. Also included are a video polarization switch and other optional features. Since there are several systems which can be purchased, the reader is urged to contact the people at Downlink in order to obtain personalized information on what is necessary for adequate reception in your area.

Comtech Data Corporation in Tempe, Arizona is another supplier of complete TVRO Earth stations which are offered in many different packages. The differences from package to package usually involve the size of the parabolic antenna included. Comtech manufactures several which are designed for optimum performance in various signal strength areas. The Comtech satellite receiver has been discussed elsewhere in this book and is incorporated into an overall system with a low noise amplifier and various connecting cables.

Comtech offers three different dish antennas of 3 meters, 4 meters, and 5 meters in diameter. The basic package includes the 3-meter dish with an EL/AZ mount. Other options include a polar mount with manual adjustment, one with motorized adjustment, and another with a programmed positioner. The same options are available with the 4-meter antenna, but the 5-meter dish is not fitted with a polar mount. Obviously, with this many antennas and mounting options, many different packages are made available to the consumer. Further information can be obtained from Comtech Antenna Corporation in St. Cloud, Florida.

Winegard is a name which has long been associated with television antennas. Their satellite communications division is now offering complete Earth stations which consist of the Winegard Model SC-7024 satellite video

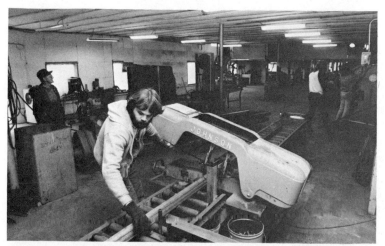

Fig. 8-3. Downlink's manufacturing facility is presently turning out over 500 spherical antenna kits per month (courtesy Downlink, Inc.).

receiver, which includes digital channel selection, and a self-contained LNA power supply. The features and specifications of this model, which is not discussed elsewhere, are given in Fig. 8-5. Winegard also offers five different low noise amplifiers which are enclosed in rugged, dip-brazed aluminum housings. The LNA which is chosen as part of a systems package will depend upon the diameter of the antenna used and the satellite signal strength in a particular area. These LNAs have noise temperatures of 85, 90, 100, 110, and 120 degrees Kelvin. All have typical gains of 50 dB minimum.

Fig. 8-4. In the Skyview II kit, the receiver is mounted near the antenna site (courtesy Downlink, Inc.).

Winegard offers two basic parabolic antennas with diameters of ten and twelve feet respectively. A prime focus feed is used with the LNAs at the focal point. This is a dual polarized feed which requires the two LNAs to be mounted on its mating flanges.

Again, there are several antenna options and five different LNAs to choose from, so system components and prices will vary greatly. But this is a complete TVRO Earth station package and the manufacturer should be able to supply interested readers with information on the specific package components which are recommended for his site area. Winegard may be contacted by writing the Winegard Company, 3000 Kirkwood Street, Burlington, Iowa, 52601.

The complete Winegard TVRO package comes with a detailed step by step instruction manual. Included is information on how to locate the satellite and what special tools are required. The entire package can be assembled by one person with the exception of mounting the antenna, which will require two or three persons. Everything is included for complete installation except the cement, which is user-supplied. The 10-foot TVRO package comes in five separate cartons and weighs 552 pounds. With the 12-foot antenna, weight is increased to 690 pounds.

This chapter has highlighted a few of the companies which offer complete TVRO Earth station packages. There are many others around and more are entering the market every month. A complete listing of these package companies is included in the appendices at the back of this book.

Complete TVRO packages are one of the most attractive satellite television buys today. This assumes that the owner does not intend to build any discrete components. The system will probably go together faster owing to the assembly manuals which are always included. These provide step-by-step instructions during every phase of construction. There is usually no problem in interconnecting the various parts, as all of the complications have been worked out in advance by the supplier. There is certainly a lot to be said for starting a project with all parts on hand. Too often, when parts are ordered from several different sources, they arrive at many different times. If a station installation is begun without all parts on hand, it may have to sit idly for several weeks awaiting the arrival of that final circuit or mounting bracket. Partially assembled projects of any type are often more susceptible to damage from the elements, so even if you do assemble a station by this latter method, don't begin until you have everything on hand.

Another important feature of complete TVRO Earth station packages is the possible cost savings which can be had. Large suppliers can buy components in quantity from small manufacturers at a much better price than you can get by buying direct. A portion of a quantity discount is often passed along to the Earth station buyer, having a direct impact on total cost. This chapter has mentioned a few specifics where up to $1,500 was saved by buying a complete package. The same will generally hold true in most package offers.

STANDARD FEATURES		De-emphasis	75 micro seconds
DC Block		Impedance	600 ohm balanced
Remote Capability		Level	0dBm ±3dB
Nom. 18VDC LNA Power		Harmonic Distortion	<1%
6.2/6.8MHz PGM Demod		Connector	Terminal Strip
LED Channel Display			
Digital Channel Select		**POWER**	
100' Interconnect Cable		Prime	117VAC ±10%, 60Hz Nom.
		Consumption	20 Watts
DOWN CONVERTER		LNA (unregulated)	Nom. 18VDC
Frequency	3.7 - 4.2GHz		(Terminal Strip)
Impedance	50 ohm		
Input Range	−60 to −30dBm	**CONTROLS**	
Nominal Return Loss	15dB	Power On/Off	
Nominal Noise Figure	15dB	AGC/MGC Select	
Threshold	8dB	Channel Select	
Connector	Type N	Local/Remote Select	
Output	70MHz	Video Level Adjust	
		Audio Level Adjust	
IF		Manual Gain Adjust	
Frequency	70MHz	6.2/6.8MHz Select	
Bandwidth	30MHz Typ		
AGC Range	30dB	**MECHANICAL**	
		Size	
TEST POINTS		Receiver Control	4" × 20" × 15" deep
AGC or MGC		Downconverter	4" × 16" × 10" deep
Signal Level	Terminal Strip	Weight	18 pounds
		Mounting	
VIDEO		Rec. Conv.	Table Top
Format	525/50 or 60	Down Converter	Antenna Frame
Frequency Response	±.5dB, 15HZ/		
	4MHz	**OPTIONS**	
Polarity	Positive White	Remote Control (50' cable)	
Dispersion Removal	>40dB	5.8/7.4 -7 / 6MHz Program Demods	
Output	1 Volt P/P ±3dB	Modulator, channel 3 or 4 selectable	
Video Out	Type F	Auto Dual Pole Switching	
Composite Out	Type F	RG-214 (20' with conn.)	
AUDIO			
Subcarrier	6.2/6.8MHz standard		
Frequency Response	±.5dB, 15Hz		
	to 15KHz		

Fig. 8-5. Specifications for the Winegard model SC-7024 satellite video receiver (courtesy Winegard Company).

There are many different routes to take when buying a TVRO Earth station. In the future, complete kits for home construction of LNAs, receivers, antennas, modulators, etc. will certainly be in abundance. Already, some of these are beginning to be offered to the average consumer. But until this occurs, for most persons, the complete TVRO Earth station component package is hard to beat.

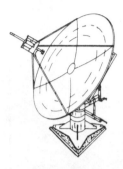

Chapter 9
Multiple Access
Ground Stations

A multiple access TVRO ground station is defined as a satellite reception system which allows many persons to tap in. This would apply to the many rooms in hotels, motels, apartments, townhouse complexes, condominiums, and even neighborhood associations. The system is basically the same as a personal Earth station but involves tying in with standard vhf and uhf antenna systems and then running the total output through a splitting system to which cables from each location are connected.

Owning and operating a multiple access satellite television system has recently become a rather unusual yet secure investment opportunity for the commercial dwelling owner or neighborhood association. It has been proven that the installation of a quality satellite receiving system can dramatically increase the occupancy rate and rental receipt of many types of multiple dwelling units. From a resale standpoint, property value and marketability is also significantly increased.

Third Wave Communications Corporation in Brighton, Michigan, a leader in the TVRO field, recently presented information about Holiday Inn and this well-known chain's experience with multiple access satellite receiving systems. Reporting in a major financial journal, Holiday Inn stated that the installation of satellite reception systems in over 100 of their inns was both financially and promotionally successful. The participating inns that have their dishes visible from the adjacent highways reported a 20% increase in occupancy rates.

The availability of closed circuit corporate teleconferencing by means of satellite has also increased rentals to the national business conference market. Apartments, condominiums, co-op, and neighborhood associations, in addition to hotel and motel owners, may find that they will not only be providing tenants or guests with the convenience and selection of the rich variety of satellite programming, but may also profit substantially both in direct cash receipts and in Federal tax credits.

While this book is aimed more at the single user, multiple access systems do not differ greatly and may be of interest to those readers who live in multiple dwelling complexes. It may be possible to form a co-op composed of your neighbors or to strike a deal with the complex owner to install such a system. Installing your own multiple access systems allows you to act as a private cable system. This differs from commercial cable companies, in that once the system is installed, you don't have to pay a middleman. If there is no local cable company in your area, the attractiveness of a multiple access satellite ground station is enhanced.

Make no mistake about it. Satellite television is of interest to nearly everyone and commercial establishments are taking advantage of this. For example, in Las Vegas, the marquis at the top of many hotels read, "Slots Pay 97%, Satellite Television!" Commercial cable companies and privately owned multiple access satellite reception systems are not a passing fad. Rather, they are a permanent part of the entertainment and communications evolution and revolution which has taken place in the United States.

According to a major electronics magazine, Third Wave Communications Corporation is one of the leaders in the manufacture and installation of privately owned and operated multiple access satellite reception systems. This company offers a very informative brochure and questionnaire which can aid the individual interested in multiple access systems in determining what system or systems to purchase.

According to Third Wave Communications, most multiple access systems and local cable systems offer several channels of satellite programming. This is in addition to a number of local television stations. A typical system might include one or two movie channels, such as Home Box Office, Showtime, and/or The Movie Channel. Also included might be Cable Network News, The Superstation (WTBS), and a sports program, such as ESPN. We're talking about quite a number of channels here, but all of them are available from one satellite named Satcom I. A personal station can easily receive the same programs that the local cable companies provide. This permits the owner of a small apartment complex to place himself in an extremely competitive position. Of course, the best part is that *you* own the system and determine the programming.

Legally, the FCC states that anyone can own and operate a TVRO Earth station. You are not required to have a license. However, if more than 50 dwellings are involved and they are not under common ownership and control, you must register as a cable company.

Figure 9-1 shows the Third Wave Communications Corporation's TVRO 6000 Multiple Access Satellite Receiving System. This is the basic layout, and the entire system can be adapted to meet specific needs and requirements. Also, additions can be made after the initial installation. These might include security systems, stereo music via satellite, or even an interface with television cameras. The system is designed for the addition of any new developments that technology produces in the coming

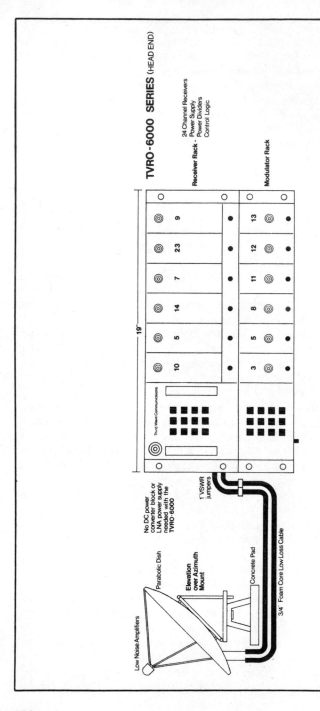

TVRO - 6000 SERIES (HEAD END)

Receiver Rack - 24 Channel Receivers
Power Supply
Power Dividers
Control Logic

Modulator Rack

19"

Third West Communications

No DC power
converter block or
LNA power supply
needed with the
TVRO - 6000

1" VSWR jumpers

Low Noise Amplifiers

Parabolic Dish

Elevation
over Azimuth
Mount

Concrete Pad

3/4" Foam Core Low Loss Cable

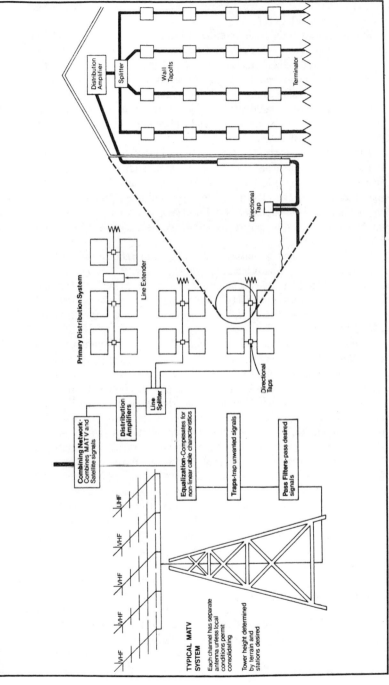

Fig. 9-1. TVRO 6000 multiple access satellite receiving system (courtesy Third Wave Communications Corporation).

187

years. The TVRO 6000 receiving system is available with as few as two or up to twenty channels.

Figure 9-2 shows the basic antenna, which is a parabolic dish. This can be mounted on ground level or on a rooftop. The ground level installation is preferred due to the extra cost usually involved in supporting or bracing the dish on a roof. Ground installation simply involves the attachment of the mount to a concrete slab. A rooftop mount would probably be necessary in dense urban areas where buildings are major obstructions. Of course, large trees or mountains are also a problem and will be more likely encountered in suburban and rural areas.

The electronics portion of this system is stored in a weatherproof container close to the dish in order to prevent signal loss and from a financial standpoint, to reduce the cost of long cable runs. The actual space required for housing the electronics receiver and modulator will depend upon the number of channels desired. For example, a six-channel receiver requires a space which is approximately 22 inches wide, 14 inches deep, and 36 inches high.

Buildings that have a master antenna system require only what is known as a head end. This principle is shown in Fig. 9-1. Here, the dish antenna is installed and cables connected to the receiver rack. The present master antenna system is connected to the modulator rack by a combining network which merges the satellite TV signals as well as those from the standard TV antennas.

Two dish antennas are available from Third Wave Communications. One is 3.66 meters in diameter, while the other has a diameter of a full 5 meters. Your geographical location will determine which antenna will be needed for good reception. Also included in the TVRO 6000 system are two low noise amplifiers, two power dividers, weatherproof electronic housing, and a receiver and modulator for each channel selected from the satellite. This is in addition to all cables for connecting the head-end to your system.

If there are enough unused vhf channels on the current TV system, one or more satellite signals can be sent to any of these open channels. Unfortunately, this would limit the expansion capabilities of the entire system, owing to the fact that there are only 12 vhf channels. The alternative to this is to modulate the satellite channel to the mid-band frequency. This would require a converter at each television set. This latter system is far more costly but allows for greater expansion in the future. The actual cost will be determined by the number of televisions attached to the system, since each will require a mid-band converter.

To take this discussion one step further, let's assume that you do not have a present master distribution system (vhf-uhf antenna installation system). It then becomes necessary to either run cables from the satellite receivers into each building's distribution system or even to combine the systems, installing a master antenna for vhf/uhf reception at the dish site and then sending the combined signals into each building. This would apply more to a neighborhood complex than to condominiums.

188

Fig. 9-2. The antenna of this multiple access system is a parabolic dish (courtesy Third Wave Communications Corporation).

From a purely financial standpoint, an extremely favorable profit picture can be drawn for many owners of personal cable systems. To repeat, it is your system and there will be no varying costs and no adjustments due to inflation. With the proper pre-planning, operating a private cable system is not only convenient; it's profitable. Third Wave Communications has included a sample picture based upon an installation in a large 150-unit complex. In some areas, rents can be raised by $25.00 per unit to accommodate the new system. 150 units at an average increase of $25.00 provides an additional $3,750 per month. Over a five-year period, this could mean an incremental growth increase of $225,000.

Of course, you have to consider the fact that you may not be at 100% occupancy at all times. Then, you should figure an average of $1,000 per year for maintenance. A six-channel system with a mid-band converter will cost $39,500. Adding $5,000 for five years of average maintenance, the total cost of $44,500. Five-year tax credit would be $2,633 the first year (investment tax credit), plus five years depreciation tax credit of $3,950 per year. This would provide a grand total of $22,383 in tax credits. Since the system costs $44,500 and the tax credit total is $22,383, your actual five-year cost is only $22,117. If you could reach the five-year incremental rent increase total of $225,000, you would have a five year profit of $202,617 based only upon the criteria in this discussion. You would need to generate an income of $4,423.40 per year in order to break even.

There are a lot of variables here, so it must be stated that these figures are provided in order to give the reader interested in a multi-access installation an idea of what *can* be accomplished with good planning and excellent management. Figure 9-3 provides a list of typical costs for the satellite head-end only systems offered by Third Wave Communications. Figure 9-4 is a reprint of the questionnaire supplied with this company's information packet to enable them to give you an accurate quote. Persons who reside in multi-dwelling units might want to get together and explore this possibility if enough are interested in personal satellite reception.

It should be pointed out that when you go the multiple access TVRO Earth station route, everyone will have to decide on which satellite is to be received. Some of the channels offered by Satcom I have been mentioned, but there are other satellites available with programming to suit a myriad of interests. Remember, when many persons are using the same antenna and distribution system, you can't go outside and turn the antenna to another

2 channels	$ 18,000.00
3 channels	22,250.00
4 channels	26,500.00
5 channels	31,000.00
6 channels	35,250.00
7 channels	39,500.00
8 channels	43,750.00
9 channels	48,000.00
10 channels	52,250.00
11 channels	56,500.00
12 channels	60,750.00
13 channels	65,000.00
14 channels	69,250.00
15 channels	73,500.00
16 channels	77,750.00
17 channels	82,000.00
18 channels	86,250.00
19 channels	90,500.00
20 channels	94,750.00
21 channels	99,000.00
22 channels	103,250.00

Fig. 9-3. Typical costs for the satellite head-end only system (courtesy Third Wave Communications Corporation).

Fill out the following information:

Address of proposed site _____

Latitude _____ Longitude _____
(Nearest major city _____)

If you already own a master antenna and distribution system for your complex, then all we need to know is how many programs you wish to receive off the satellite. (See attached list.)
_____
_____

If you do not have a master antenna system already (new projects included), then complete the following:

1. How many channels do you wish to receive off the satellite? (See attached list.)
_____
2. Send layout of your complex showing each building, roads, underground utilities and accurate distances between all buildings and roads.
3. Number of apartments in each building. _____
4. Does each building have its own television distribution system? ___Yes ___No
5. If yes, please describe the system:

   a) Type of cable used.
      1. Coaxial cable type _____
      2. Other _____
   b) Are there existing splitters and amplifiers in each building?
      1. Type of splitter and quantity
         2-way ___   4-way ___   6-way ___   8-way ___
         Other _____
      2. Type of distribution amplifier and quantity
         Name _____
         Model number _____
   c) If you are unable to answer the previous questions, please explain why
_____
_____
_____
_____

6. Please check what other services you would like included in the system:

      ___ FM radio antenna
      ___ Fire alarm
      ___ Burglar alarm
      ___ Camera capability
      ___ Other
   a) What building or area do you want the camera in?
      ___ Teletext capability
7. Please list all the local TV stations that you would like to receive, including their channel numbers and the cities from which they originate.
_____
_____
_____
_____
_____
_____
_____
_____

Fig. 9-4. Third Wave's questionnaire is helpful in providing them with information to enable them to estimate the cost for a system (courtesy Third Wave Communications Corporation).

satellite just because what's presently playing is not what you want to see. Almost all multiple access Earth stations are equipped with dish antennas which are relatively fixed as to position. You must choose a satellite with programming which is in most demand by the majority of users.

Now, suppose you live in a small subdivision which contains only nine or ten homes. It might be possible for all homeowners to get together, form a co-op (relating to the Earth station only), and purchase a system. Due to the generally spread-out nature of this type of subdivision, it would probably be necessary to contract for the cable runs that would be required. Assuming the cost of a six-channel system to be approximately $40,000, with perhaps another $10,000 thrown in for extras, the entire per-dwelling price (based upon ten homes) would be $5,000. With average maintenance of $1,000 per year, an additional $100 per year would be required from each home as well.

After the first year, the system is completely paid for and all future maintenance is automatically shared by each home. The value of each residence should increase due to the personally owned system. Ownership would automatically be transferred as homes were sold to other individuals and would most likely be made a part of the deed. Maintenance fees would be paid directly to the co-op or owner's association. Before contemplating this type of installation, be sure to check with local officials as to zoning and any other requirements which might be necessary. It might be a good idea, too, to mutually hire an attorney to handle the legal aspects of writing an agreement or covenant which all owners can live with and which will be attractive to potential buyers of these homes in the future.

This chapter has deviated a bit from the overall subject of this book, which is the installation, owning, and operation of personal Earth stations. The multiaccess system described is personal, in that only one person or a small group of persons owns the system, but is removed from our basic subject by the fact that there may be many users.

Homeowners, apartment dwellers, etc. may want to explore multiple access systems with the idea of possibly saving money by sharing the total cost. Unfortunately, most multiple access systems are priced far higher than personal Earth stations due to the commercial quality of most components and to the additional equipment, such as splitters, combining networks, etc. Persons who think they may be able to save by going the multiple access method are urged to explore this type of system as well as the true personal Earth station. When all the facts are in, an intelligent decision may be made which is in the best economic interests of all concerned.

Chapter 10
Satellite
Location Techniques

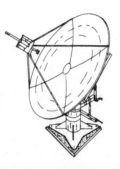

Once you have arrived at a design for your Earth station, it is absolutely essential to decide which satellite or satellites you desire to receive signals from. This would have been a simple process a decade or so ago before the communications satellites began appearing en mass over the Earth. Today, there are many, many satellites which broadcast television signals back to Earth.

All TV satellites are put into special orbits over the Earth which cause them to hover motionless in the sky. They are located some 22,000 miles above the surface of the Earth and rotate in sync with our home planet. As far as we on the Earth are concerned, the satellites are motionless because they rotate with the Earth. These are called geostationary or geosynchronous satellites due to their orbits, which are stationary or synchronous with the Earth's rotation. All of the entertainment satellites we are interested in are located high above the equator, but each one is stationed at a different longitude. Various satellites are positioned in very exacting orbits in order for their transmissions to cover different parts of the Earth.

In order to receive TV signals from these satellites, it is necessary to use a dish-shaped antenna which is aimed directly at the satellite whose signals are to be received. If the aiming is even a hair off, the received signal will be very poor or nonexistent. This assumes that all Earth station equipment is properly set up and functioning as it was designed to do.

There are many geostationary satellites and the ones we're interested in broadcast special programming. Home Box Office, Showtime, Superstation, and The Movie Channel are some examples of what can be received in the entertainment line from these special satellites. Other services include cable and pay TV programs, uncut feature-length movies, sports, educaional and religious programs, and news. There are also geostationary communications satellites which carry long distance and overseas telephone conversations. Other orbiting bodies include navigation satellites, weather satellites, and even military spy satellites. From the TV satellites alone, there are presently more than 75 channels available to viewers in North and South America and almost as many in other parts of the world.

While these entertainment satellites broadcast signals which may be received by viewers in North and South America, this does not mean that all viewers can receive them. Your particular location on the globe will determine which satellites can be effectively "hit" by the antenna. The terrain immediately surrounding the Earth station location will also be a determining factor and applies especially to the quality of reception. Obviously, if your receiving antenna is located on one side of a mountain and the orbiting satellite is on the other side, it would be necessary for the transmitted signal from space to travel through the mountain in order to be received. This is not possible.

As is the case with most antennas, the greatest reception efficiency will be had from systems which mount the parabolic dish antenna in a high, clear location. This should also provide the capability of receiving several satellites by simply changing the position of the dish antenna. In the above example, the top of the mountain would have been the most ideal location for the antenna, but losses between the antenna and the receiver would be quite prohibitive, since a sizeable travel distance would be required.

When using simple antennas such as dipoles and verticals for shortwave communications work, it is usually a simple matter to mount the elements fairly high in the air, thus avoiding nearby obstructions. This is not the case with Earth station antennas, which can be described as massive in size as compared with the shortwave antenna. To mount one of the microwave dishes high in the sky would require a massive tower which could easily cost ten times as much as all of your Earth station equipment. Also, signals received from an orbiting satellite are line-of-sight. This means that the signal travels directly from the satellite to the Earth without any appreciable bending or skip. Shortwave signals often bounce from the Earth to the ionosphere and then back again. They may do this many times in a single transmission; thus, it is possible to communicate with receiving stations which are completely blocked from the transmitting antenna by the curvature of the Earth.

Fortunately for all persons interested in their own personal Earth stations, there are many satellites in equatorial orbits. In most situations, at least one and probably many of these will transmit signals which can be received at your location, regardless of where it might be. It is necessary to find out what is available in your area before you can accurately determine your station location.

This is easily accomplished and for a minimal fee by obtaining a computer printout of the satellites which can be received from your area. Each printout will be different for various station locations across the

United States. You will find that some satellites are completely "hidden" from your location, while others will serve as prime reception targets for your antenna.

How do you obtain a computer printout? In many instances, the manufacturers of complete Earth stations will supply one for your particular location when you express an interest in purchasing some of their products. This is not usually done free of charge, although some companies will refund the price of the computer search if you actually make a purchase. For example, Heathkit offers a full site survey packet which has full information on their Earth station and will help determine if your site is acceptable for installation. The price is currently $30.00 and you receive a computer printout which shows satellite "look angles" for your address, a compass for determining the directions of different satellites, and an inclinometer to help determine the degree of elevation between the Earth's surface and the satellite to assure a clear line-of-sight path for reception. Upon receiving this kit, you can make a good determination as to which satellites may be received in your area and, using the inclinometer, you can also find out if there are any major local obstructions such as mountains, towers, etc. in the signal path. Should you decide to order Heathkit's Earth station, the $30.00 Site Survey kit price is deducted from the equipment price. This example of obtaining a site survey is typical of the many reputable manufacturers of satellite Earth stations.

A less expensive route to go when you simply desire a computer printout for your area is offered by Satellite Computer Service, 1808 Pomona Drive, Las Cruces, New Mexico, 88001. This company offers an independent computer analysis of your location, anywhere in the world. Their computer will tell you which satellites can be received at your location, where to point your antenna, and how strong the signals are likely to be. The cost for this service is $19.95 (add $2.00 extra if you want it sent by air mail). I wrote to this firm and was able to obtain a computer printout for my area. This printout is shown in Fig. 10-1A. In looking over the specifications provided, they at first appear to be rather complex. A closer examination shows that the information is very basic and easily understood by even the most non-technical individual. Satellite longitude, elevation, azimuth, and loss in decibels at various frequencies are provided in a clear, understandable manner. The computer even indicates which satellite longitudinal locations cannot be received. The computer printout is used with a list of geostationary satellite longitudes and frequencies, which is also provided by Satellite Computer Service. See Fig. 10-1B. This allows the user to determine from the list and computer printout combination which satellites are available to his or her particular area.

It can be seen from the printout that only a very few (seven) satellite locations would not be generally usable for the author's location in the Southeast. Satellite Computer Service determines your geographic coordinates from your address and gives them to the computer. It then uses these numbers to solve a series of complex mathematical equations involving

SAT LON	AZ	EL	LOSS IN DB AT 1 GHZ	4 GHZ	8 GHZ	12 GHZ
160 W	***	**	***	***	***	***
159 W	***	**	***	***	***	***
158 W	***	**	***	***	***	***
157 W	***	**	***	***	***	***
156 W	262	1	186	199	207	211
155 W	262	2	186	199	207	211
154 W	261	2	186	199	206	211
153 W	260	3	186	199	206	211
152 W	260	4	186	199	206	211
151 W	259	5	186	199	206	211
150 W	258	5	186	199	206	211
149 W	258	6	186	199	206	211
148 W	257	7	186	199	206	211
147 W	256	8	186	199	206	211
146 W	256	8	186	199	206	211
145 W	255	9	186	199	206	211
144 W	254	10	186	199	206	211
143 W	254	11	186	199	206	211
142 W	253	12	186	199	206	211
141 W	252	12	186	199	206	211
140 W	251	13	186	199	206	211
139 W	251	14	186	199	206	211
138 W	250	15	186	199	206	211
137 W	249	15	186	199	206	211
136 W	248	16	186	199	206	211
135 W	248	17	186	199	206	211
134 W	247	18	186	199	206	211
133 W	246	18	186	199	206	211
132 W	245	19	186	199	206	210
131 W	245	20	186	198	206	210
130 W	244	21	186	198	206	210
129 W	243	21	185	198	206	210
128 W	242	22	185	198	206	210
127 W	241	23	185	198	206	210
126 W	240	24	185	198	205	210
125 W	239	24	185	198	205	210
124 W	239	25	185	198	205	210
123 W	238	26	185	198	205	210
122 W	237	26	185	198	205	210
121 W	236	27	185	198	205	210
120 W	235	28	185	198	205	210

Fig. 10-1A. A computer printout will provide you with information regarding satellite reception for your area (courtesy Satellite Computer Service).

			LOSS	IN	DB	AT
SAT			1	4	8	12
LON	AZ	EL	GHZ	GHZ	GHZ	GHZ
119 W	234	28	185	198	205	210
118 W	233	29	185	198	205	210
117 W	232	30	185	198	205	210
116 W	231	30	185	198	205	210
115 W	230	31	185	198	205	209
114 W	229	32	185	198	205	209
113 W	228	32	185	198	205	209
112 W	227	33	185	198	205	209
111 W	226	34	185	198	205	209
110 W	225	34	185	198	205	209
109 W	224	35	185	198	205	209
108 W	222	35	185	198	205	209
107 W	221	36	185	198	205	209
106 W	220	37	185	198	205	209
105 W	219	37	185	198	204	209
104 W	218	38	185	198	204	209
103 W	216	38	185	198	204	209
102 W	215	39	185	197	204	209
101 W	214	39	185	197	204	209
100 W	213	40	185	197	204	209
99 W	211	40	185	197	204	209
98 W	210	40	185	197	204	209
97 W	208	41	185	197	204	208
96 W	207	41	185	197	204	208
95 W	206	42	185	197	204	208
94 W	204	42	185	197	204	208
93 W	203	42	185	197	204	208
92 W	201	43	185	197	204	208
91 W	200	43	185	197	204	208
90 W	198	43	185	197	204	208
89 W	197	44	185	197	204	208
88 W	195	44	185	197	204	208
87 W	194	44	185	197	204	208
86 W	192	44	185	197	204	208
85 W	191	44	185	197	204	208
84 W	189	45	185	197	204	208
83 W	188	45	185	197	204	208
82 W	186	45	185	197	204	208
81 W	184	45	185	197	204	208
80 W	183	45	185	197	204	208
79 W	181	45	185	197	204	208

*** MEANS THAT A GEOSTATIONARY SATELLITE AT THE
GIVEN POSITION IS NOT VISIBLE FROM THIS GROUND
STATION. SATELLITES AT LONGITUDES NOT LISTED ARE
NOT VISIBLE FROM THIS GROUND STATION.

SAT LON		AZ	EL	LOSS IN DB AT			
				1 GHZ	4 GHZ	8 GHZ	12 GHZ
78	W	180	45	185	197	204	208
77	W	178	45	185	197	204	208
76	W	177	45	185	197	204	208
75	W	175	45	185	197	204	208
74	W	173	45	185	197	204	208
73	W	172	45	185	197	204	208
72	W	170	44	185	197	204	208
71	W	169	44	185	197	204	208
70	W	167	44	185	197	204	208
69	W	166	44	185	197	204	208
68	W	164	44	185	197	204	208
67	W	163	43	185	197	204	208
66	W	161	43	185	197	204	208
65	W	160	43	185	197	204	208
64	W	158	43	185	197	204	208
63	W	157	42	185	197	204	208
62	W	155	42	185	197	204	208
61	W	154	42	185	197	204	208
60	W	152	41	185	197	204	208
59	W	151	41	185	197	204	208
58	W	150	40	185	197	204	209
57	W	148	40	185	197	204	209
56	W	147	39	185	197	204	209
55	W	146	39	185	197	204	209
54	W	144	38	185	198	204	209
53	W	143	38	185	198	204	209
52	W	142	37	185	198	204	209
51	W	141	37	185	198	204	209
50	W	140	36	185	198	205	209
49	W	138	36	185	198	205	209
48	W	137	35	185	198	205	209
47	W	136	35	185	198	205	209
46	W	135	34	185	198	205	209
45	W	134	33	185	198	205	209
44	W	133	33	185	198	205	209
43	W	132	32	185	198	205	209
42	W	131	32	185	198	205	209
41	W	130	31	185	198	205	209
40	W	129	30	185	198	205	210
39	W	128	30	185	198	205	210
38	W	127	29	185	198	205	210

Fig. 10-1A. A computer printout will provide you with information regarding satellite reception for your area (courtesy Satellite Computer Service). (Continued from page 196.)

| | | | LOSS | IN | DB | AT |
| SAT | | | 1 | 4 | 8 | 12 |
LON	AZ	EL	GHZ	GHZ	GHZ	GHZ
37 W	126	28	185	198	205	210
36 W	125	28	185	198	205	210
35 W	124	27	185	198	205	210
34 W	123	26	185	198	205	210
33 W	122	25	185	198	205	210
32 W	121	25	185	198	205	210
31 W	120	24	185	198	205	210
30 W	119	23	185	198	205	210
29 W	118	23	185	198	206	210
28 W	118	22	185	198	206	210
27 W	117	21	186	198	206	210
26 W	116	20	186	198	206	210
25 W	115	20	186	198	206	210
24 W	114	19	186	199	206	210
23 W	114	18	186	199	206	211
22 W	113	17	186	199	206	211
21 W	112	17	186	199	206	211
20 W	111	16	186	199	206	211
19 W	111	15	186	199	206	211
18 W	110	14	186	199	206	211
17 W	109	14	186	199	206	211
16 W	108	13	186	199	206	211
15 W	108	12	186	199	206	211
14 W	107	11	186	199	206	211
13 W	106	11	186	199	206	211
12 W	105	10	186	199	206	211
11 W	105	9	186	199	206	211
10 W	104	8	186	199	206	211
9 W	103	7	186	199	206	211
8 W	103	7	186	199	206	211
7 W	102	6	186	199	206	211
6 W	101	5	186	199	206	211
5 W	101	4	186	199	206	211
4 W	100	4	186	199	206	211
3 W	99	3	186	199	206	211
2 W	99	2	186	199	207	211
1 W	98	1	186	199	207	211
0 E	97	0	186	199	207	211
1 E	***	**	***	***	***	***
2 E	***	**	***	***	***	***
3 E	***	**	***	***	***	***

List of Geostationary Satellite Longitudes and Frequencies

SAT LON	NAME	OWNER	USE	FREQ GHz
0 W	Meteosat 1	ESA	Weather	0.136, 1.7
0 W	Meteosat 2	ESA	Weather	0.136, 1.7
1 W	INTELSAT-IV-E	INTELSAT	TV/Tel	4, 11
2.6 W	INTELSAT-IV-A	INTELSAT	TV/Tel	4, 11
4 W	INTELSAT-IV-F2	INTELSAT	TV/Tel	4, 11
11.6 W	Symphonie-2	France / W. Ger.	Exp.	4
13 W	DSCS-2-F7	USA	Mil.	7
13.5	Statsionar-4	USSR	TV/Tel	4
14 W	Volna-2	USSR	Commo	1.5
14 W	Loutch-1	USSR	TV/Tel	11
14.2 W	Ghorizout-2	India	TV/Tel	4, 11
15W	Marisat	COMSAT	Maritime	0.25
15 W	SIRIO-1	Italy	Exp./Met.	0.136
18 W	NATO-3A	NATO	Mil.	2.2, 7
18.5 W	INTELSAT-IV-F1	INTELSAT	TV/Tel.	4
21.5 W	INTELSAT-IV-F3	INTELSAT	TV/Tel.	4
23 W	FLTSATCOM-3	USA	Mil.	0.25, 7
24.6 W	INTELSAT-IV-A-F1	INTELSAT	TV/Tel.	4
25 W	Volna-1	USSR	Commo	0.28, 1.5
25 W	Stasionar-8	USSR	TV/Tel.	4
25 W	Loutch-P1	USSR	TV/Tel.	11
25 W	Gals-1	USSR	Mil.	7
27.5 W	INTELSAT-IV-A-F2	INTELSAT	TV/Tel.	4
34.5 W	INTELSAT-IV-A-F4	INTELSAT	TV/Tel.	4
41 W	TDRS	USA-NASA	Commo	2.2, 13
44 W	LES-9	USA	Exp.	7, 32
50 W	NATO-3C	NATO	Mil.	2.2, 7
70 W	FLTSATCOM	USA	Mil.	0.25, 7
70 W	ATS-5	USA	Exp.	0.136
75 W	Hughes-1	USA	TV/Tel.	4
75 W	SMS-2	USA	Weather	0.136, 1.7
85 W	LES-6	USA	Research	—
86.9 W	Comstar-D3	COMSAT	TV/Tel	4, 20, 30
90 W	GOES-1	USA	Weather	0, 136, 1.7
90.9 W	Westar-3	WU-USA	TV/Tel	4
95 W	Comstar-2	COMSAT	TV/Tel	4, 18, 28
99 W	TDRS	WU-USA	Commo	4, 12, 13
99 W	Westar 1	WU-USA	TV/Tel	4
100 W	LES-9	USA	Exp.	7, 30
100 W	FLTSATCOM-1	USA	Mil.	0.25, 7
103.9 W	ANIK-A1	Canada	TV/Tel.	4
105.2 W	ATS-3	USA	Exp.	0.136, 4
106 W	SBS-B	USA	Tel / Commo	12
107 W	GOES-2	USA	Weather	0.136, 1.7
108.9 W	ANIK-B1	Canada	TV/Tel.	4, 12
110 W	LES-8	USA	Exp.	7, 30
112.5	ANIK-C1	Canada	TV/Tel.	12
113 W	SMS-1	USA	Weather	0.136. 1.7
113.9 W	ANIK-A3	Canada	TV/Tel.	4
114 W	ANIK-A2	Canada	TV/Tel.	4, 12
116 W	ANIK-C2	Canada	TV/Tel.	12
116 W	CTS	USA/Canada	Exp.	12
118.9 W	SATCOM-2-F1	RCA-USA	TV/Tel	4
119 W	SBS-B	USA	Tel /Commo	12
122 W	SBS-A	USA	Tel / Commo	12

Fig. 10-1B. List of geostationary satellite longitudes and frequencies (courtesy Satellite Computer Service).

SAT LON	NAME	OWNER	USE	FREQ GHz
123.5 W	Westar 2	WU-USA	TV/Tel	4
127.8	Comstar-D1	COMSAT	TV/Tel	4, 18, 28
130 W	DSCS-2-F9	USA	Mil.	2, 7
132 W	SATCOM III	RCA-USA	TV/Tel.	4
134.9	SATCOM-1-F1	RCA-USA	TV/Tel.	4
135 W	GOES-3	USA	Weather	0.136, 1.7
135 W	GOES-4	USA	Weather	0.136, 1.7
135 W	SMS-2	USA	Weather	0.136, 1.7
135 W	DSCS-2-F11	USA	Mil.	2, 7
135 W	DSCS-2-F14	USA	Mil.	2, 7
135 W	NATO-3B-F2	NATO	Mil.	2.3, 7
140 W	ATS-6	USA	Exp.	1.5, 4, 20, 30
149 W	ATS-1	USA	Exp.	0.136
170 W	Volna-7	USSR	Commo	0.25, 1.5
170 W	Statsionar-10	USSR	TV/Tel.	4
170 W	Loutch-P4	USSR	Mil.	11
170 W	Gals-4	USSR	Mil.	7
171 W	TDRS	WU-USA	Commo	2.2, 13
0 E	Nordsat	Nordic Nations	TV/Tel.	12
10 E	OTS-2	ESA	Exp.	0.138, 11
35 E	Raduga-3	USSR	TV/Tel.	4
35 E	Statsionar-2A	USSR	TV/Tel.	4
35 E	Raduga-4	USSR	TV/Tel.	4
40 E	Marecs-A	ESA	Maritime	1, 5, 4
45 E	Loutch-P2	USSR	TV/Tel.	11
45 E	Statsionar-9	USSR	TV/Tel.	4
45 E	Gals-2	USSR	Mil.	7
45 E	Volna-3	USSR	Commo	0.25, 1.5
49 E	Symphonie-1	France/W. Ger.	Exp.	4
53 E	Ekran-3	USSR	TV/Tel.	0.714
53 E	Ekran-4	USSR	TV/Tel.	0.714
53 E	Gorizout-3	USSR	Mil.	4, 7
53 E	Loutch-2	USSR	TV/Mil.	11
54 E	DSCS-2-F4	USA	Mil.	8
56.5 E	INTELSAT-III-F3	INTELSAT	TV/Tel.	4
60 E	INTELSAT-IV-A-F6	INTELSAT	TV/Tel.	4
60.2 E	INTELSAT-IV-A-F5	INTELSAT	TV/Tel.	4
61.4 E	INTELSAT-IV-F1	INTELSAT	TV/Tel.	4
63 E	INTELSAT-IV-A-F3	INTELSAT	TV/Tel.	4
70 E	GOMS	USSR	Weather	0.136, 1.7
70 E	STW-2	P. R. China	Exp./Mil.	4
71 E	Insat	India	Exp.	2.5, 4
73 E	Marisat	COMSAT	Maritime	0.25
74 E	Insat-1A	India	TV/Tel.	2.5, 4
75 E	FLTSATCOM	USA	Mil.	2.2, 7
77 E	Palapa-2	Indonesia	TV/Tel.	4
80 E	Statsionar-1B	USSR	TV/Tel.	4
80 E	Raduga-1	USSR	TV/Tel.	4
80 E	Raduga-2	USSR	TV/Tel.	4
83 E	Palapa-A1	Indonesia	TV/Tel.	4
85 E	Gals-3	USSR	TV/Tel.	7
85 E	Loutch-P3	USSR	TV/Tel.	11
85 E	Raduga-2	USSR	TV/Tel.	4
85 E	Voina-5	USSR	Commo	0.25, 1.5
90 E	Loutch-3	USSR	TV/Tel.	11

SAT LON	NAME	OWNER	USE	FREQ GHz
90 E	Statsionar-6	USSr	TV/Tel.	4
99 E	Statsionar-T	USSR	TV	0.714
99 E	Ekran-1	USSR	TV	0.714
99 E	Ekran-2	USSR	TV	0.714
110 E	BSE	Japan	TV	12
125 E	STW-1	P. R. China	Exp./Mil.	4
130 E	KIKU-2	Japan	Exp.	0.136, 1.7, 11
135 E	CSE	Japan	Exp.	4, 20
135 E	ECS-2	Japan	TV/Tel.	4, 32
140 E	GMS-1	Japan	Weather	0.137, 1.7
140 E	GMS-2	Japan	Weather	0.137, 1.7
140 E	Loutch-4	USSR	Mil.	11
140 E	Statsionar-7	USSR	TV/Tel.	4
140 E	Volna-6	USSR	Commo	1.5
145 E	ECS	Japan	Exp.	4, 32
172 E	FLTSATCOM-4	USA	Mil.	2.2, 7
174 E	INTELSAT-IV-F8	INTELSAT	TV/Tel.	4
175 E	DSCS-2-F8	USA	Mil.	7
175 E	DSCS-2-F10	USA	Mil.	7
175 E	DSCS-2-F12	USA	Mil.	7
175 E	DSCS-2-F13	USA	Mil.	7
176.5 E	Marisat	COMSAT	Maritime	0.25, 1.5, 4
179 E	INTELSAT-IV-F4	INTELSAT	TV/Tel.	4

Abbreviations

Commo.	VHF-UHF aircraft and mobile communications, Telemetry, digital data, etc.
Exp.	Experimental, misc. research.
TV/Tel.	Television, Telephone, miscellaneous communications.
Mil.	Military communications.
Maritime	Shipboard navigation and communication.
Notes:	Not all satellites listed are active at this time. Some are on standby status for various operational reasons. This list is believed to be accurate at the time of publication. Nevertheless, the user should be aware that satellites are occasionally relocated, and new ones are placed into operation. Information about such changes is published in a number of aviation and space journals.
	Portions of this material were derived, with permission, from the Comsat Technical Review, Volume 10, Number 1, "Geosynchronous Satellite Log for 1980," by W. L. Morgan.

Fig. 10-1B. List of geostationary satellite longitudes and frequencies (courtesy Satellite Computer Service). (Continued from page 200.)

spherical trigonometry and logarithms. It is worth mentioning here that this particular mathematical analysis does not just cover TV satellites, but also every possible geostationary satellite location and does so in one-degree increments of longitude.

The first step in the computer's process of evaluating your location is to determine which satellite longitudes are not "visible" from your location. These are then eliminated from the printout. Obviously, the word visible is used loosely here to describe those satellites which are below the horizon from a location, plus making it impossible to receive transmissions fr̶ them.

The satellite longitudes which are visible from your location are the̶ listed on the computer printout, along with some very important numbers. These numbers will tell you where to look in the sky to "see" each satellite from your location. They also give an indication of how strong or weak the signals from each particular satellite are likely to be at your location.

To better understand how to use the printout to determine which satellites you will be able to receive transmissions from, the first step is to refer to the previously mentioned List of Geostationary Satellite Longitudes and Frequencies (Fig. 10-1B) which Satellite Computer Service includes as part of their package. On the list, identify a satellite that you are interested in. For example, let's use the printout provided for the author and assume that we wish to receive SATCOM I, which is a television satellite owned by RCA and offers several popular movie channels. By referring to the listing of longitudes and frequencies, we can determine that this particular satellite's longitude is 134.9 W. Make a note of this figure, making sure to pay careful attention to whether the longitude is east (E) or west (W). If the indicated longitude contains a decimal part, as in the case of our example, round it off to the nearest whole number. This would make the longitudinal location of the SATCOM I satellite 135W. Also, make a note of the satellite frequency, which in our case is 4 GHz.

The next step is to go to the computer printout (Fig. 10-1A) and look up the same satellite longitude, paying careful attention to whether the longitude is east or west. Referring to Fig. 10-1, it can be seen that 135 W does appear on the printout, which gives us a preliminary indication that we should be able to receive signals from the SATCOM I satellite. However, if a particular longitude did not appear on the printout, this means that we would not be able to "see" that satellite and therefore would not be able to receive signals from it.

Now that we know that the longitudinal position of the SATCOM I satellite makes it possible for us to receive its signals, some additional checks are in order. Notice on the printout that the next two columns of numbers after the satellite longitude are labelled AZ and EL. These are the azimuth and elevation numbers which tell the user exactly which direction to look in order to "see" the satellite. Before explaining how to use these numbers, however, it is important to discuss some information that should be kept in mind during this process.

When using the term "looking" with regard to "seeing" a satellite, this refers to the direction that the antenna would be aimed in order to point at a particular satellite. Of course, you can't really see the satellite with your eyes, since most are located more than 22,000 miles away. Nevertheless, in order to make sure that you can receive signals from the satellite, you will have to stand in the location that you plan to mount your antenna (in the backyard, on the roof, etc.) and actually look in the direction that the computer printout provides.

The reason for doing this is that your satellite TV antenna must have absolutely clear shot at the satellite. There cannot be obstructions of any ᴵnd in the way. Hills, buildings, trees, houses, or even a telephone pole will completely block out the signal. Therefore, even if the computer printout shows that you may be able to receive signals from a particular satellite, you must make sure that there are no local obstructions that would prevent you from doing so. Usually, such problems can be overcome by relocating your antenna site. In the case of nearby hills or large buildings, however, you may not be able to move far enough to be able to use a particular satellite.

Also, a good rule of thumb to follow is that if the listed elevation (EL) figure is 10 or less, the signal will most likely be of poor quality even if the antenna has a clear view of the satellite. This is due to the fact that the antenna will be pointing so close to the surface of the Earth that it will pick up a great deal of man-made interference. Unless you can afford to spend an additional amount of money on some quite expensive equipment, it is wise to avoid those signals from a satellite which have an elevation figure of less than 10.

Another point to keep in mind is that some satellites have highly directional transmitting antennas which concentrate their signals toward very specific areas. In cases such as this, you may find that you are within "view" of the satellite, but you might not be within its strongest area of coverage. Some of the ANIK satellites, for example, beam their signals primarily toward the northern provinces of Canada. In this example, a viewer in Canada would probably receive much stronger signals than would a viewer in Mexico, even though the distance from the satellite is farther and the elevation angle is lower.

With this information in mind, let's return to the process of determining whether or not we will be able to receive a good signal from the SATCOM I satellite at my location. Both azimuth (AZ) and elevation (EL) angles, or directions, which are measured in degrees. The AZ figure provides the compass direction, from 0 to 360 degrees, measured clockwise from true north. Refer to Fig. 10-2 for a simplified diagram of compass directions. In this view, we are looking down on a man who is facing north (0 degrees AZ). The AZ figures are drawn around him. It can be seen that South is an AZ of 180 degrees, North-East an AZ of 45 degrees, and so forth. The azimuth figure for the SATCOM I satellite, which is at a longitude of 140 W, is 251 degrees. In other words, we must first face true north (0 degrees azimuth) and then, using a compass, determine the location of 251 degrees azimuth and face in that direction.

Keep in mind that the azimuth figures are based on true north, which in most locations is somewhat different from magnetic north. If you don't know how to locate true north directly, you can use an inexpensive dime-store compass to locate magnetic north, but it will be necessary to convert the printout AZ figure to a magnetic azimuth. This is done by referring to Fig. 10-3 and marking your approximate position on the map. In the case of

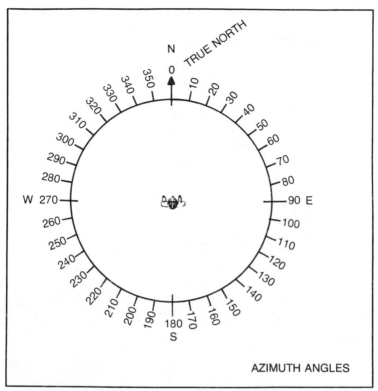

Fig. 10-2. Azimuth angles, indicating true north (courtesy Satellite Computer Service).

my location, which is in Virginia, it would be between the dotted lines indicating 4 degrees and 8 degrees. These dotted lines indicate the amount of magnetic variation in the different parts of the country. Now, follow the dotted line which is closest to my location, which is the 8 degree line, out to the end. This 8 degree figure will be used to estimate the amount of magnetic variation for my area. It is also important to make note that this 8 degree variation is westerly (W), as indicated on the map. In our example, since the variation is westerly, the 8 degree figure is added to the computer printout AZ figure of 251, or 251 + 8 = 259. Therefore, 259 would be the magnetic AZ figure to use. However, if the variation were easterly, the indicated amount would be subtracted from the computer printout AZ figure.

To illustrate further, suppose you live in Denver, Colorado, and the satellite that you're interested in has a computer printout AZ figure of 163. Referring to Fig. 10-3, it can be seen that the magnetic variation at Denver is about 13 east (E). Thus, this amount is subtracted from the printout AZ, or 163 - 13 = 150. Therefore, 150 would be the magnetic AZ figure to use.

Fig. 10-3. The dotted lines on this map indicate magnetic variation (courtesy Satellite Computer Service).

In yet another example, suppose you live in Erie, Pennsylvania and the satellite that you're interested in has a computer printout AZ figure of 191. Again referring to Fig. 10-3, the magnetic variation at Erie is 7 west (W). Therefore, this amount is added to the printout AZ, or 191 + 7 = 198. The figure of 198 AZ would thus be the magnetic AZ figure to use.

If you happen to live outside the United States, it will be necessary to obtain the magnetic variation figures for your particular area from another source. This information is generally available from either a public library or possibly a nearby airport.

Now that we have determined which direction it will be necessary to face, the next step is to find out the elevation at which to look. Elevation (EL) angles are measured from the horizontal (0 degrees) to the vertical (90 degrees). In other words, if you look straight ahead toward the horizon, this is an elevation of 0. If you tilt your head back until you are looking straight up, the elevation is 90. Although the computer printout does provided the exact elevation angle, most persons will find it a bit difficult to estimate this angle by sight alone. However, this can be determined using a few simple materials with a much greater degree of accuracy.

Shown in Fig. 10-4 is a protractor, which can be used to measure elevation angles. The protractor shown here can be traced and then cut out, or you can purchase an inexpensive plastic protractor from an office supply store. Now, referring to Fig. 10-5, use the protractor to mark off an angle equal to the satellite elevation angle, which in our example is 13, on a piece of cardboard. A triangle is then cut from the cardboard as indicated by the mark made using the protractor. Be sure to mark the letters "EL" on the corner of the triangle that you measured with the protractor so that you won't forget which corner of the triangle to refer to.

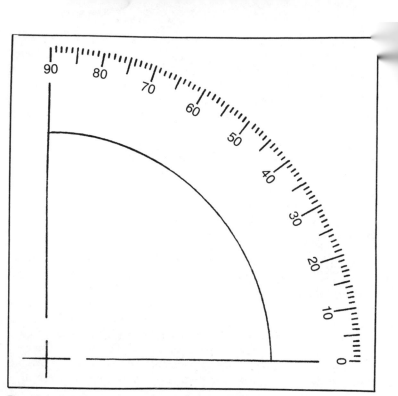

Fig. 10-4. A protractor is used to make a cardboard angle of your location's elevation (courtesy Satellite Computer Service).

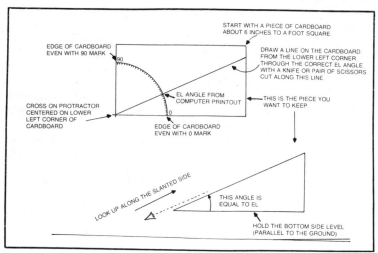

Fig. 10-5. Using the protractor, the proper elevation is marked off (courtesy Satellite Computer Service).

Now, stand at the location which has been selected for mounting of the antenna. Face the direction of the satellite's azimuth, which in our example is 259 and hold the cardboard triangle even with your eyes, keeping the bottom side horizontal (parallel to the ground). Sight up the slope of the triangle as shown in Fig. 10-5. You should now be looking directly toward the satellite, although you already know that it won't be possible to actually see it.

What we are looking for at this point is any type of obstruction that might block the satellite's signals. If, when sighting along the slope of the triangle, your view is blocked by obstructions of any kind, such as trees, hills, houses, etc., then you will not be able to receive signals from the satellite. If, however, all you see in this direction is the sky, you should be able to receive this particular satellite's transmission. In the case of my location when performing this procedure there was a mountain range in the vicinity, but it did appear to be approximately 4 or 5 degrees below our proposed antenna aiming point. If it was found that there were some obstructions that would prevent reception of a particular satellite, it would have been necessary to either change the location of the antenna or possibly provide some means of evaluating it at the original proposed site.

This completes the procedure for determining whether or not you will be able to receive signals from a particular satellite. Of course, this process will have to be repeated for each additional satellite that you may be interested in. This will involve preparing a new cardboard elevation triangle each time.

Once you have selected a satellite and determined that you can receive signals from it at your particular location, the final step will be to once again refer to the computer printout in order to check the strength or weakness of the signals. Remember that when originally checking the computer printout, we made a note of the satellite frequency. Again referring to the printout, notice that there are four columns of numbers after the figures for SAT LON, AZ, and EL. These numbers are labeled "LOSS IN DB AT 1 GHz, 4 GHz, 8 GHz, and 12 GHz."

Go to the column which most closely corresponds to the frequency noted earlier, which in our example is 4 GHz. However, if your frequency was 5 GHz, for example, you would go to the column labeled 4 GHz also. In that column (taking care to stay on the same line as the SAT LON that was used earlier to find the azimuth and elevation), you will find a number which indicates how strong or weak the signal will be for your location. The lower the dB LOSS number, the stronger the TV signal will be. The higher this number is, the weaker the TV signal will be. The lowest dB LOSS number that will be encountered on the printout will be about 183. This indicates the best (strongest) signals. The highest dB LOSS number will be about 215, which indicates the worst (weakest) signals. The weaker signals will require a larger antenna and a more expensive converter to receive them with any degree of quality. Since the frequency for my location is 186, this is a good indication that the signals from the SATCOM I satellite can be received quite strongly at the proposed location for my earth station.

It should be kept in mind when requesting a computer printout for a proposed Earth station site that a single printout is calculated for only one specific location. Any additional proposed sites will require a separate printout. Also, if the requested printout does not include a particular geostationary satellite's longitude (SAT LON), this means that you cannot receive transmissions from that satellite at your proposed location. If a satellite's longitude does appear on the printout but its elevation figure is less than 10, you might be able to receive signals from it, but they will probably be of poor quality. Take care when setting up your TVRO antenna that it is aimed directly at the satellite you wish to receive transmissions from using the azimuth and elevation figures as a guide and making sure that there are no obstructions between the antenna and the satellite toward which it is pointing. Following the guidelines provided here and installing your Earth station properly, you should be able to receive quality transmissions from your antenna arrangement.

It is necessary to reemphasize the fact that site selection will play an important part in the reception of satellite signals. Just because you have a "window" to an entertainment satellite in your area does not necessarily mean that you can receive it from any site in the same vicinity. As was previously mentioned, local obstructions, especially those lying close by the Earth station, can effectively close that window and make reception impossible. In these situations, moving the Earth station antenna to a slightly different location may open the window fully. Another chapter in this book deals more fully with the selection of Earth station sites.

It is mandatory that all persons wishing to install their own satellite Earth station obtain the data and perform the calculations necessary to determine which satellites beam signals that are receivable at a particular site. If a satellite you desire to receive is not available in your area, there is nothing you can do to correct this situation short of moving to an area which has an open window. This will undoubtedly be many miles (or several hundred) from your home location. Even the best ground station equipment available will do nothing to open a window which does not exist.

Fortunately, there are many satellites available to all areas of North and South America, and the enterprising Earth station builder will certainly find one or more which broadcast the programs desired.

Satellite location data should be obtained *before* you decide to actually purchase the components to build the Earth station. This is especially true if you desire to receive signals from one particular satellite. If there is no window to this satellite in your area, then you may decide to forego the project entirely. This chapter has overviewed the methods used to obtain computer printouts and to properly determine the direction, elevation and azimuth for the antenna installation. Most manufacturers of satellite ground station equipment are in a position to supply this information (usually for a fee) or to advise you as to where this information can be obtained.

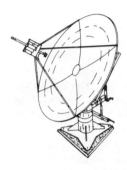

Chapter 11

A Commercial
Cable Television System

Previous chapters in this book have described personal Earth stations and their application to home use. Many areas in the United States offer cable television systems which supply service to customers who access their facilities. Many potential buyers of personal Earth stations will wonder about the differences in their system when compared with a commercial installation and what advantages are to be gained by the latter.

In order to present a good idea of what is involved in a commercial installation, I contacted Teleco, Inc., a cable TV company which serves the Front Royal, Virginia area. Teleco, Inc. is owned by Mid Atlantic and has been in business for several decades. Only recently has satellite TV been added to the cable, which formerly offered only vhf and uhf channels. the vhf/uhf antenna site was located several miles from Front Royal, Virginia. While this was an excellent site for this type of reception, it was totally unsuitable for an Earth station due to mountains which would not allow a clear window on Satcom I, the satellite which supplies Home Box Office, ESPN, The Super Station, and other programs which were to be offered.

An in-town location was obtained for the building of the Earth station. Fortunately, this was only a couple of hundred feet away from my offices. The manager of Teleco, Inc. was quite cooperative in explaining the operation of his company's system and even gave a personal tour of the facility, pointing out each piece of equipment and phase of operation in the order in which they are used.

The first difference noted between commercial installations and personal Earth stations was the fact that the parabolic dish antenna is much larger (see Fig. 11-1). Most dish antennas used for personal TVRO Earth stations are ten to twelve feet in diameter. The Teleco, Inc. dish is a full five

Fig. 11-1. The Teleco, Inc. commercial Earth station's dish measures a full five meters or sixteen feet wide (courtesy Teleco, Inc.).

meters (sixteen feet) wide and presented a massive appearance upon close examination. Unlike most personal Earth stations, this antenna was not readily adjustable in regard to elevation and azimuth. A more or less permanent mounting technique was used, which bolts everything into place. Elevation and azimuth can be adjusted, but not in a short period of time. This requires unbolting several connections and/or manually "horsing" the antenna base into the desired position. Of course, easy manual adjustment is not necessary in commercial installations, because the antenna, once adjusted, is left in the fixed position permanently. The manager pointed out that it would soon be necessary to make a major readjustment in the antenna's position because of a new satellite which will be used by the same programmer. This satellite will take over the handling of the same programs presently offered by Satcom I. Often, satellite users will switch as the orbit of one satellite in use begins to decay. Commercial Earth stations require excellent windows. These deteriorate directly in relationship to time in orbit. When the programmer leaves this satellite, other services will begin leasing it. These latter uses are not as critical as commercial television reception, so the satellite will still be used for a long time to come.

The new satellite which will be accessed by Teleco, Inc. lies at a slightly different point on the horizon. When it comes time to switch, the antenna will simply be readjusted and will function as before, but from a new satellite.

Another difference can be readily discerned by looking at the feed horn/LNA assembly at the focal point. Whereas most personal Earth stations have a single LNA here, the commercial station has two. Satellite signals are transmitted either as horizontally polarized waves or those which are vertically polarized. Most personal Earth stations use a single LNA and feed horn but incorporate a standard antenna rotor which will allow the assembly to be turned 90°. If the system is set up to receive horizontally polarized signals, the rotor, when activated, will perform the 90° shift, which converts the system to vertical polarization. Since an LNA can easily cost $800 or more, it is far more cost efficient to use an inexpensive rotor system rather than two low noise amplifiers. You might wonder why the commercial facility doesn't do the same thing. There's a good reason. It just won't work. At your personal station, when you must switch from vertical to horizontal polarization, you do it from a remote control located atop your TV. But remember, you and your family are the only users of this personal system and only one channel can be selected at a time. In a commercial facility, there are thousands of users. Some of them will be tuned to one channel offered by the cable company, while others will simultaneously be receiving another available channel. Suppose the first channel is transmitted by vertical polarization, while the other is horizontally polarized. If a single LNA is in the vertically polarized position, then those persons wishing to receive the horizontal channel will be out of luck.

Unlike personal Earth stations, cable companies offer several satellite channels simultaneously. This requires a different receiver for each channel and the capability (at the antenna) for simultaneous reception of both horizontally and vertically polarized transmissions.

Since two separate feed horn/LNA combinations are used at the antenna site, there are two transmission lines back to the bank of receivers. The line from the horizontal LNA is attached to receivers set up for satellite transponders which transmit horizontally polarized signals. The same is true of the other line, which feeds signals to receivers set up on vertically polarized channels. Figure 11-2 shows the LNA/feed horn assembly at the antenna focal point.

At this point, we can say that one of the major differences in cable systems as opposed to personal Earth stations is the fact that several

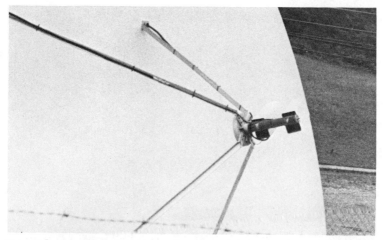

Fig. 11-2. The Teleco low noise amplifier (LNA)/feed horn assembly at the antenna focal point.

simultaneous channels are offered. This makes the entire system far more complex, as it is necessary to add (in addition to the two LNAs) more receivers and modulators, one of each for every channel. This is one of the reasons why cable television Earth stations usually cost well over $100,000.

Figure 11-3 shows the instrumentation rack which is enclosed in a small building at the Earth station site. Contained here are the receivers, modulators, monitors, and some amplifiers for the Earth station. Additionally, the combining network which brings in the vhf/uhf channels from the other antenna site is found at this location. This latter network performs basically the same function as was described for combining signals in an earlier chapter on multiple access Earth stations.

The cable used to bring the signals the short distance from the antenna to the electronic apparatus is a special low-loss coaxial type. This is shown in Fig. 11-4. Connections are made to the back of the panel shown earlier.

Before discussing the bank of receivers and modulators, it is important to make another distinction here between equipment intended for general public uses and that which is intended for commercial purposes. The latter is usually built to much closer tolerances and is generally more rugged and dependable than the former. Commercial receivers and modulators are designed for untended operation. Your personal Earth station equipment is not. Also, one must consider the duty factor. Few personal Earth stations will be operated 24 hours a day, 365 days a year. Most cable companies, on the other hand, offer full time operation day in and day out. Obviously, their equipment must be designed to withstand this amount of usage. While circuits of equipment intended for the two purposes (personal and commercial) may seem theoretically comparable, the actual choice of components

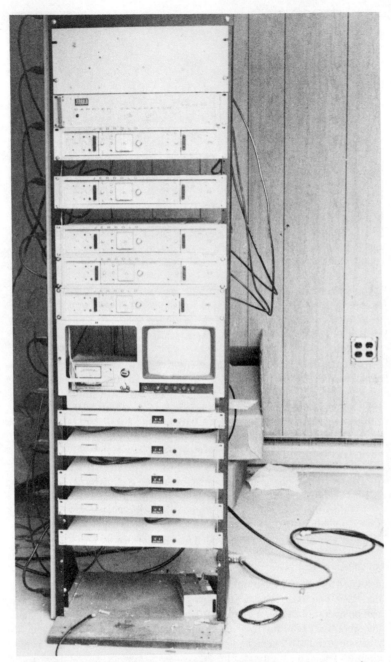

Fig. 11-3. The instrumentation rack contains receivers, modulators, monitors, and amplifiers.

214

Fig. 11-4. The cable used in this commercial installation is a special low-loss coaxial type.

used will differ greatly. Commercial electronic equipment usually is designed with larger power supplies and with commercial quality components which will provide a longer and more dependable service life. These are often called commercial grade or military grade components.

Obviously, the choice of select and often specialized circuits and components has to cost more. The cable company is paying extra for this equipment because it needs more services from its devices than you would as a personal Earth station owner. Commercial receivers can cost more than an entire personal Earth station facility. And remember, the cable company may need six or more, depending upon the number of channels it is to offer its customers. A different modulator will be needed for each channel as well. The modulators used by Teleco, Inc. cost in excess of $3,000 each.

The receiver and modulator act in the same manner as those incorporated in personal Earth station installations. The video and audio output from the detected microwave signals are superimposed on the carrier generated by the modulator. Another difference makes itself known at this point. Before discussing this, a review of the portion of the cable company Earth station presented thus far reflects a lot of equipment, installation time, and cost. And remember, we still haven't gotten the signal to the home television receiver yet.

The difference which occurs at this point when comparing cable Earth stations with those intended for personal use includes the massive distribution system which must be run to every customer's home. However, the modulator at the antenna site does not usually have an output at a frequency which can be received directly by the standard television set. Usually, the output frequency of the modulator lies above the vhf television band and below the uhf band. Why is this done? First of all, if the output were on one of the vhf channels, this might mean the dropping of one presently received station for every satellite channel offered. Today, cable companies often present a range of programs on all channels 2 through 13. Secondly, cable companies almost always offer satellite television reception as an option to customers who are already on the cable and are presently receiving the vhf channels. Customers have the option of paying an additional fee each month for the satellite channels; but if they don't want them, they can continue to receive vhf channels and pay no more.

In order for the cable company to offer satellite reception as an option, they purchase modulators which produce frequencies that fall outside of the normal television bands. These signals are on the line which attaches to your television receiver, but you can't detect them without a converter. The cable company installs a converter at each home which accepts the modulator output and changes it to a frequency within the television band. This is usually somewhere in the uhf channel area. Here is another major expense for the cable company, because if 6,000 customers want satellite reception, 6,000 converters are necessary, one at each set.

Since their inception, cable television companies have had problems with some disreputable persons stealing their signals. In other words, a few persons who did not subscribe to the company found ways of intercepting their output. This was most often done by tapping into one of the coaxial lines.

Most of the stealing occurred in homes which subscribed to the service and who then attached more television sets to the single incoming line. Cable companies usually charge a minimal fee for each additional television in a home and must use splitters which are company-installed to provide for the additional outputs.

Like anything else, if there is a way to get something for nothing, even though it may be against the law, some persons will jump to take advantage of the situation. The converters which the cable companies use to provide satellite channel reception in each home are available to the general public through special electronic outlets. Most are reputable firms, but a few specialize in offering equipment specifically for the individual who wants to beat the system. These companies often advertise in questionable manners. To prevent this situation from becoming widespread, many cable companies will install frequency traps between the main signal line and any home which is taking cable service. When the subscriber signs up for satellite reception, the technician who installs the converter also removes the frequency trap. This latter device effectively blocks the satellite signals from the homes which do not subscribe to the optional service. It is unfortunate that these companies find it necessary to go to all this trouble, but statistics would prove that it's absolutely essential. Less than one-half of one percent of the subscribers would even consider doing anything like stealing from a cable company, but the company has no way of knowing exactly who comprises the dishonest minority. All customers pay for their potential dishonesty by mutually footing the bill for all those frequency traps, their cost being passed along to the subscribers through increased service rates.

A converter unit is installed by the cable company at each satellite subscriber's set. This is a type of modulator which accepts the input signal from the cable and then converts it to a channel within the uhf television band.

Another problem is often encountered by cable companies in that some subscribers still have older television models which do not offer uhf tuning. Here, a special converter may be used which allows one vhf channel to be used for reception of all the satellite channels. This converter has its own channel knob which selects all of the satellite programs being offered by the company. The output of this unit may be on any vhf channel 2 through 13 and there is a cutout switch so that the vhf channel may also be used to receive the vhf station which is replaced by the satellite signals during that mode of operation.

In addition to all of the equipment necessary to operate a successful cable TV company, the redundancy factor must be considered. Most commercial equipment has a bit of redundancy built into it, but most companies still keep a spare receiver, modulator, LNA, and other vital equipment on hand in case of emergency. Even the finest equipment can fail unexpectedly and cable customers get mad when they can't receive what they're paying for. Redundancy is another necessary expense of cable company operation.

While there are many differences in a cable company operation, let's now examine the similarities between it and a personal TVRO Earth station. Both use dish antennas, LNAs, receivers and modulators. Both receive signals in the same manner and from the same satellites. Both ultimately offer outputs which are designed to be received by standard television sets. The only major differences include backup devices, redundancy equipment, special tuning arrangements, cost, and size.

Cable companies must obtain a stronger signal than is necessary for practical personal Earth station uses. A lot of equipment is attached to the output of the antenna and this tends to lessen the level of the output to each receiver. Also, the cable company must provide constant and dependable reception during all weather conditions. For example, a large cable company's parabolic dish antenna is far more firmly seated on its mounting site than would be a typical personal Earth station antenna. Both will probably have similar wind survival ratings, but the commercial installation will be far less affected as to reception by moderately high winds. The personal Earth station antenna will probably induce a bit of flutter and fading when winds gust to over 50 miles per hour. This won't occur in the system discussed in this chapter. Also, during periods of heavy rain, high temperatures, snow and ice, the increased signal strength provided by the large commercial dish will allow the cable company to provide a signal to users which appears no different than those received during optimum weather conditions. The commercial antenna actually provides more gain than is needed most of the time. But the backup is needed in order to assure that the same quality of signal is available even under the most adverse conditions. This will not be true of most personal Earth stations. Ninety percent of the time, the received image and audio on a television receiver operated from a personal Earth station will be as good as that which is provided by the cable company. But you will notice some difference when adverse weather conditions are in effect. Sure, you could purchase a larger antenna and LNAs with better noise figures, but you're talking about doubling or tripling the price of your personal Earth station. This is simply not cost effective for the small advantage these extras provide.

It can be seen that while the principles involved in satellite television reception are basically the same in cable company installations as well as in personal TVRO Earth stations, the demands placed upon the commercial station are far more numerous. Both fields are closely allied and the TVRO experimenter is urged to be on the lookout for commercial surplus equipment as it becomes available. In some instances, it is more profitable for a cable company to simply replace malfunctioning equipment which is out of warranty than to have it repaired. These units are then relegated to a dusty shelf in a back room and can often be purchased for next to nothing. A visit to your local cable company may result in making acquaintances which will aid you in your future personal TVRO Earth station effort.

I would like to thank Teleco, Inc. and especially its manager, Richard Burke, for the assistance offered in researching this chapter. I think the reader will find that most cable company personnel are more than willing to give advice and lend experience in addressing problems which can develop with personal Earth stations. Actually, the two fields are a bit competitive, but there is little chance that in the near future, personal Earth stations will cause any significant financial losses to the cable companies. Satellite television reception is still a new field. Most persons in this area of endeavor are pleased by the experimentation which is being conducted by major companies, manufacturers, and individuals. This is something we all have in common.

Chapter 12
Building Your Own

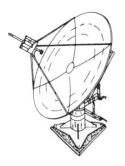

As is the case with any electronic field of endeavor, there are those persons who are not content to use only commercially manufactured equipment. These are the experimenters, the people who like to build as much of their equipment as possible. In doing so, they gain a better understanding of how each system operates and are better able to perform repairs and, more importantly, to make modifications which can result in vast system improvements.

Unfortunately, the TVRO field is still in its infancy. There is not a great deal of technical information available which is produced in a form that can be readily understood and applied by the average home experimenter. The avid home builder can take heart in the fact that electronics kits are now being offered by a few companies which will allow you to build TVRO Earth station components such as the LNA, downverter, receiver, etc. As of this writing, most of these kits are designed more for experimental purposes than useful applications, although some systems will be able to make immediate use of them with the proper interfaces.

One such company is Gillaspie & Associates, Inc., 177 Webster Street, Suite A455, Monterey, California, 93949. The Gillaspie engineering team is made up of two prominent microwave authorities. Norman Gillaspie has had an impressive career in video circuitry, having been involved in television broadcasting for eleven years. He worked for the ABC Television Network Engineering Department and has written articles for a number of electronics magazines.

Werner Vavken is the other authority and is known for his work in satellite systems technology. He has worked for over ten years for large microwave companies and has been a consultant for many of them. Mr.

Vavken heads up the Gillaspie research and development projects which are an integral part of the operation.

Gillaspie offers primarily completed TVRO receivers, the newest of which is fully synthesized and provides video and audio scan. They also offer several auxiliary components. These are discussed elsewhere in this book.

Another individual with Gillaspie & Associates is Harry Harp, who heads up the business and marketing end. Harry quite readily admits he is a non-technical member of this company.

Of prime interest to the discussion in this chapter is the fact that Gillaspie & Associates also offers three kits which are designed for the home experimenter. In talking with Harry Harp, I learned that Mr. Gillaspie is an amateur radio operator and likes the idea of persons building whatever they can. This helps to advance the state of the art. In complying with his feelings, three solid-state kits are presently offered by this company which are designed to be assembled by persons with average electronic experience. Harry Harp told me that even he had assembled one of the kits and it worked just fine. At the same time, he pointed out that the kits being offered are designed for the experimenter and were not intended to match the performance of many types of commercial equipment being offered today. He went on to say, however, that these kits can serve as building blocks for a working TVRO installation in many cases.

The price of the various kits seems to be very reasonable. For example, the Gillaspie GaAs FET Low Noise Amplifier Kit sells for a total of $37.50. This includes the printed circuit board, seven special chip capacitors, and two SMA connectors. This is not a complete kit. It will still be necessary for the builder to purchase the expensive transistors ($55.00 each) and a few standard components such as capacitors and diodes. All told, the complete kit of parts should cost the builder less than $175. This compares with $800 or more for a commercially built LNA.

The following is a brief description of the LNA which is reprinted from information supplied by Gillaspie & Associates:

Figure 12-1 shows a pictorial diagram of the component layout, along with some important building notes. Figure 12-2 shows the LNA printed circuit board. This is depicted in actual size.

Noise figures as low as 1.2-1.3 dB have been achieved with various devices and with some tuning, using small pieces of copper soldered in various places or by removing some circuitry. For the sake of simplicity and stability, this circuit board layout seems to be a good compromise when compared to the other designs which are available to the experimenter.

The input and output vswr is quite good in this design, allowing the use of a waveguide-to-coax transition coupled with a waveguide horn to give a real-world antenna feed without the use of a lossy isolator.

Input matching has been restricted to the best compromise between input match for power transfer and optimum noise figure. Output matching to the input of the second stage GaAs FET is the simplest way possible

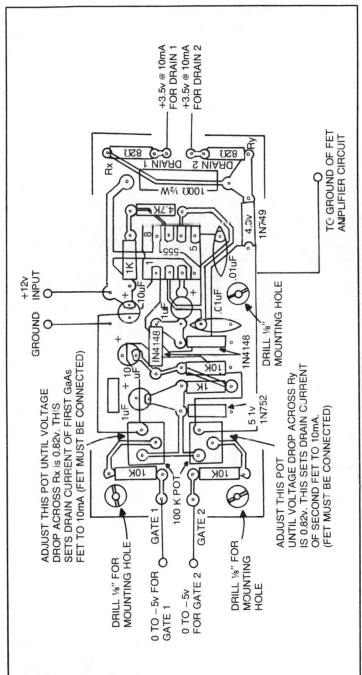

Fig. 12-1. Component layout and building notes for the Gillaspie LNA kit (courtesy Gillaspie & Associates, Inc.).

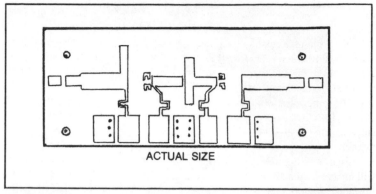

Fig. 12-2. Printed circuit board for the LNA (courtesy Gillaspie & Associates, Inc.).

using a short section of transmission line, as is the case with the output matching. The 5.6 pF coupling capacitors are chosen for low coupling loss and dc blocking. The other capacitors provide for microwave and low frequency bypassing. The 4.7 V zener diodes are strictly overvoltage protection, and 1N4148s are simple "idiot reverse voltage protection" to save your expensive GaAs FETs.

The 10 k pots are used to adjust the GaAs FET's gate voltage in order to control the drain current to 10 mA, which is the optimum for best noise figure. The procedure for setting the drain current is as follows:

■ Before applying the drain voltage, set the 10 k pots for maximum voltage at the gates.

■ Monitor the voltage drop across the 33 ohm resistor feeding the first GaAs FET.

■ Turn on the +4.5 V supply (drain voltage).

■ Adjust the 10 k pot feeding the first gate for 0.33 volts drop across the 33 ohm resistor. The current for the first GaAs FET is now set to 10 mA.

■ Now, simply monitor the voltage across the second 33 ohm resistor and adjust the second 10 k pot until the voltage across the second 33 ohm resistor is also 0.33 volts. The current to the second device is now 10 mA also.

Construction is straightforward, as the parts placement guide shown in Fig. 12-1 illustrates, but a few comments should be made at this point concerning GaAs FETs:

■ Don't wear static-causing clothing (wool, etc.)

■ Use a grounded tip soldering iron.

In order to achieve the performance as indicated, it is imperative that the GaAs FET's source leads be poked through the holes provided and soldered to the ground side of the printed circuit board. The remaining twelve holes in the circuit board should have a #14 wire poked through and soldered on both sides before attaching the associated components.

SMA connectors should be used to launch in and out of this amplifier. Care should be exercised to insure good grounding of the SMA to the ground of the circuit board. In addition, the circuit board should be either mounted in a machined enclosure or made rigid by soldering a heavy gauge wire (#16 or so) to the bottom side of the circuit board from the input to the output. Two pieces will make it quite rigid. This will prevent the fragile chip capacitors (and the GaAs FET) from breaking if the circuit board is flexed.

Figure 12-3 shows the specifications of the completed kit, and these will further be discussed in order to make a comparison with a commercial low noise amplifier. It is only fair in making this comparison to state again that we are putting up a $175 device against one in the thousand dollar range. The commercial low noise amplifier used for comparison purposes has the same frequency range and a noise temperature of 120° k. First of all, it's necessary to convert the noise temperature designation into a noise figure which can be compared with the Gillaspie & Associates kit. A 120° k noise temperature figures out to a noise figure of 1.5 dB. This is the same noise figure which is typical of our LNA kit. As was mentioned earlier, frequency range is the same. However, our inexpensive kit only has a gain of 20 dB minimum, whereas the commercial model used as a typical comparison has a gain minimum of 50 dB. This is quite a difference and put in layman's terms, the output power from the commercial LNA to the receiver is approximately equal to many times that delivered by the kit. Each time you double output power, there is a 3 dB increase. In other words, a transmitter which puts out 1 watt will increase its gain by 3 dB if the power is raised to 2 watts. But to get another 3 dB gain, we must go to twice the second figure, or 4 watts. Of course, we are talking about extremely low power levels in this application which are measured in tiny fractions of a watt.

The Gillaspie LNA kit is designed for cascade operation. This simply means installing two or more amplifiers in series with the antenna and receiver. The gain of one of these kits is 20 dB, but two of them in cascade would give you approximately 40 dB. I originally thought this would also double the noise figure, but this is not true. The noise figure will stay about the same, as determined by the first LNA. So, with two of the Gillaspie kits in series, you would have the equivalent of a single low-noise amplifier

Frequency Range:	3.7-4.2 GHz
Noise Figure:	1.5dB typical
Gain:	20dB minimum

Fig. 12-3. Complete specifications for the LNA (courtesy Gillaspie & Associates, Inc.).

with a gain of approximately 40 dB and a noise temperature of approximately 120°K. You could even insert a third LNA kit and obtain approximately 60 dB, which would be equivalent to most commercial low-noise amplifiers in regard to gain and noise temperature. Of course, a three-unit combination would cost well over $500, so you're fast approaching the price of a commercial model. The gain of the antenna you use, signal strength in your area, satellite receiver sensitivity, and other criteria will determine how effective this LNA kit will be. As was mentioned in the description of the circuit, lower noise figures have been obtained with some slight modifications to the kit. Additional experimentation in this area could bring about some very practical results using a two-unit cascade configuration.

I am not being critical of this fine kit in any way. After all, it's cheap when compared with a commercial unit and certainly would have some limited applications in its present form. With some tinkering, testing, and tears (the three T's of home building), it might be possible to arrive at a final product which, in cascade, would provide ratings which compare favorably to some commercial units and at half the price.

Another kit which is offered by Gillaspie & Associates is a bias board for the FED amplifier just described. Shown in Fig. 12-4, this miniature circuit board, when properly outfitted with about thirty components, results in a simple GaAs FET bias circuit which performs two important functions. First, it creates a regulated +3.5 volt supply for the "drains" of the FET. Secondly, it creates an adjustable negative voltage (0 to −5 Vdc) for the FET "gates".

All of this is achieved from a single +12 V power source. The positive 3.5 V supply is derived from the 12 V input by means of a simple zener diode and a dropping resistor. The negative voltage is generated by the use of a 555 IC timer and its associated circuitry. The circuit is shown in Fig. 12-5. The 555 is wired as an astable oscillator operating near 10 kilohertz. Its

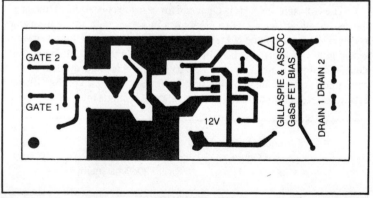

Fig. 12-4. Circuit board for the Gillaspie bias board kit (courtesy Gillaspie & Associates, Inc.).

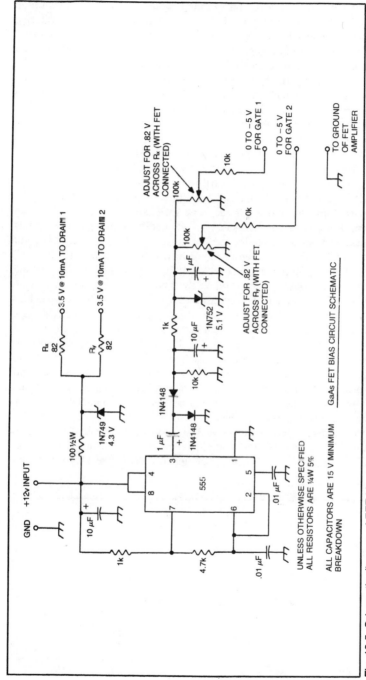

Fig. 12-5. Schematic diagram of FET bias circuit (courtesy Gillaspie & Associates, Inc.).

output is relayed to pin 3 and is rectified by the 1N4148 diode and filtered with the 10 k ohm resistor and 10 microfarad capacitor. A further filtering circuit is provided by the 1 k ohm resistor and the 1 microfarad capacitor. A zener diode serves to regulate the output to 5.1 Vdc. The 100 k ohm potentiometers provide for adjustment of this negative voltage to the desired level required by the FET gates. The 10 k ohm resistor in series with each potentiometer is there for protection. If one FET gate would become short-circuited, it will have little or no effect on the bias at the other transistors. Figure 12-6 shows the parts list. All of these are relatively common components which may be picked up at most hobby stores. These are wired to the circuit board using a small soldering pencil.

The circuit checkout procedure is very simple. First of all, you apply 12 volts dc to the bias circuit. With a sensitive voltmeter, the outputs marked "drain 1" and "drain 2" are measured. Under no load conditions, the output voltage should read 4.3 Vdc. Now, gate 1 and gate 2 are checked with the same voltmeter, again under no load conditions (the power supply is activated but *not* connected to the low noise amplifier). The readings will vary from 0 to − 5.1 Vdc when adjusting the two 100 k ohm potentiometers.

If the steps outlined so far check out, the 12 volt supply is removed and the bias circuit outputs attached to the LNA. The supply is once again activated. Now, while monitoring the voltage drop across R_x shown in Fig. 12-5, adjust the 100 k ohm potentiometer for gate 1 to 0.82 Vdc. This sets the current in the first FET to 10 milliamperes. The voltmeter is now used to monitor the voltage drop across R_y in Fig. 12-5, and the other 100 k ohm potentiometer is adjusted to give a reading of 0.82 Vdc also. The current in the second FET has not been set to the 10 milliampere value. That's all there is to it. Your complete LNA and power supply circuit are operational.

While Gillaspie & Associates does not make a complete kit of parts for a satellite receiver, they do offer the Hi-Tek Single Conversion Mixer for 3.7 to 4.2 gigahertz. The mixer is designed to get the microwave signal down to 70 megahertz with good efficiency and outstanding performance. Shown in Fig. 12-7, this stripline mixer functions with typical specifications of:

Ga As FET Bias Circuit-Parts List

Resistors	Potentiometers	Dodes
2 ea. 82 ohms ¼ W 5%	2 ea. 100K ohm Mod 63P	2 ea. 1n4148
2 ea. 1K ¼W 5%		1 oy. 1N752 (5.1v)
1 oy. 4.7K ¼W 5%	**Capacitors**	1 oy. 1N749 (4.3v.)
3 ea. 10K ¼W 5%	2 ea. .01µF Cer. Disc.	
1 oy. 100 ohms ½W 5%	2 ea. 1µF Cer. Disc. or electrolytic	**Integrated Circuits**
	2 ea. 10µF electrolytic	1 oy. 555 Timer

Fig. 12-6. Parts list for FET bias circuit (courtesy Gillaspie & Associates, Inc.).

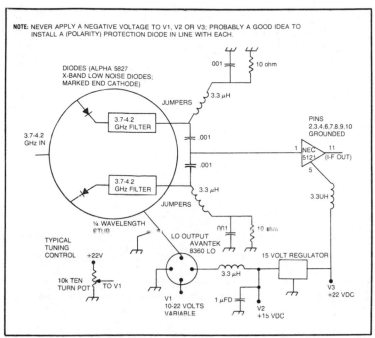

Fig. 12-7. Schematic diagram of Hi-Tek single conversion mixer (courtesy Gillaspie & Associates, Inc.).

■ Input 3.7/4.2 gigahertz, flat to plus or minus 1 dB.

■ Output at a center frequency of 70 megahertz (plus or minus 15 megahertz) with a gain variation of no more than plus or minus 0.5 dB.

■ Total conversion gain of 20 dB minimum.

Shown in pictorial form in Fig. 12-8, the stripline circuit is constructed on Diclad 527 teflon material (on 0.031 thick dielectric with one ounce copper on both sides). No substitution of this circuit board material is recommended due to the critical nature of the circuit design for these microwave frequencies.

Actually, there are three circuits on the board. These include a mixer, an oscillator, and a 30 dB gain block amplifier. For the purpose of this discussion, each section of this board will be separately discussed.

Referring to Fig. 12-8, the printed circuit line marked "rf" is the connection from the low noise amplifier to the mixer and SMA fitting is soldered to the bottom of the board as well as to the top. The signal is connected to a 1½ wavelength line which is designed to cancel the LO (local oscillator) signal at the input and output ports and to reinforce the signal at the mixer diode ports. The twin diode mixers and their placement around the ring are the other two ports. Notice the "boomerang shaped" twin inductors inside the ring. This is a 4 gigahertz low-pass filter. Also note the

227

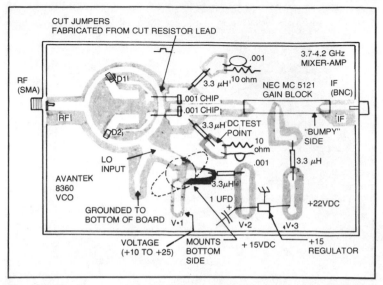

Fig. 12-8. Single conversion mixer circuit in pictorial form (courtesy Gillaspie & Associates, Inc.).

thinner etched conductors which lead to a pair of pads that couple through wire jumpers to the output line. This line combines the output from both diodes. Also attached on this line is a 3.3 microhenry choke in series with a 0.001 microfarad capacitor paralleled by a 10 ohm resistor. The network forms a low-pass filter and produces a convenient test point to determine if the local oscillator is working or whether you have damaged a diode.

Local oscillator injection comes into the classic retrace range by means of the remaining connection marked LO. This connection has a quarter wavelength line attached to it. This serves to short the ground plane on the bottom of the board. This is the dc return for the mixer diodes.

The next portion of this circuit is the vco (voltage controlled oscillator). The one recommended for this circuit is the Avantek 8360 which is shown in Fig. 12-8 and requires a 15 volt dc supply for oscillator operation plus a variable 11 to 22 volt dc supply for tuning the local oscillator. *Caution:* Never apply a negative voltage to any of the 8360 terminals.

The final portion of this circuit is a 75 ohm input/output gain block which requires 22 Vdc. There is some gain reduction when the voltage is reduced to +15 Vdc, but it is noted that the overall amount of reduction is not too severe. This entire circuit section could be powered from a 24 Vdc line with appropriate dropping resistors. The output is to the extreme right of the board, and a connection similar to the input utilizes either an SMA or BNC fitting. The operation of the unit is shown in Fig. 12-9.

Construction is very simple. All you need do is follow the parts layout. Be sure to observe polarity when connecting the diodes and the vco. All parts mount to the top of the special board with the exception of the local oscillator, which is mounted to the bottom. If you choose, the vco can be soldered to the ground plane of the board rather than using mounting hardware, although this would increase the problems associated with replacement. Additionally, the case would be heated to a higher temperature by the soldering iron. Whenever it is necessary to directly solder integrated circuits to connections, always use appropriate clipon heat sinks. The NEC gain block is the last component to be mounted. It is installed with the input side to the left and the output to the right. All unused holes in the circuit board are soldered through with short pieces of wire leads. This is done to connect the top and bottom of the ground plane together. This is a very simple circuit to assemble and should present even the beginning builder with little difficulty.

Operation of your completed circuit is even simpler than the building process. Apply the supply voltages to points V2 and V3 on the board and adjust the tuning voltage applied to the Avantek 8360 local oscillator to approximately 17 volts dc. This should place you in mid-band near 3.95 gigahertz. As per the system block diagram shown in Fig. 12-9, you would have around 52 dB of gain (electronic +40 dB passive antenna gain) up to the output of this mixer portion of the system.

To go the rest of the way, you need to pass the output of this single conversion mixer into an appropriate 20 to 30 megahertz wide i-f passband filter and then into an additional i-f gain stage of at least 30 dB. This should properly launder the signal to make it suitable for input to a phase-lock loop demodulator.

In the single conversion receiver, the local oscillator and the mixer receive the selected satellite transponder (found by varying the tuning voltage on the 8369) and convert it down to the 70 megahertz i-f. All single

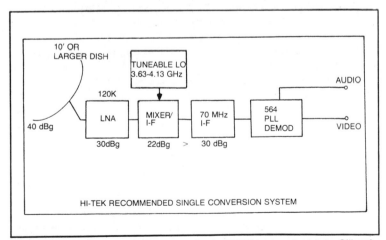

Fig. 12-9. Block diagram of mixer operation in TVRO system (courtesy Gillaspie & Associates, Inc.).

conversion mixers have some degree of image problems and a phase-lock loop demodulator minimizes this effect. The Hi-Tek Down Converter is designed so that no detectable amount of signal distortion is present in the conversion process.

One of the more appealing parts of this design is found in the fact that this system could be used to drive directly into higher i-f in the 88 to 100 megahertz region. This would allow the use of a standard FM receiver as an i-f. Using this system, you could recover audio-only transmissions where they are transmitted as discrete FM carriers by the satellite. Due to the design, the output range is unfiltered at the input to the receiver. The only additional cost would be for the FM receiver.

The Hi-Tek single conversion mixer is a good build-it-yourself circuit which will serve as a building block for a satellite receiver. There are limitations placed upon the basic single conversion design, but the home experimenter should be able to obtain pleasing results.

A TVRO RECEIVER KIT

A lot of TVRO enthusiasts will be glad to hear that a complete TVRO receiver kit is being offered by KLM Electronics, Inc., P.O. Box 816, Morgan Hill, California, 95037. Called the Sky Eye II, this is the same receiver that was mentioned in an earlier chapter. The completely assembled version sells for about $1,500. But when you order the kit version, the price drops to about $700, and you do a lot of the assembly yourself. This amounts to a savings of about $800, dropping the cost of the receiver to below that of many low noise amplifiers.

The Sky Eye II receiver is a two-part system consisting of the control console and the receiver unit. The compact control console puts satellite channel and audio tuning near the television receiver or video tape recorder. This is the only part of the receiver which is mounted in the home. The main receiver unit mounts at or near the dish antenna.

All the channels available from the satellite can be tuned within the full rotation of the channel knob. It is calibrated to the 24 channels found on Satcom I, the most popular entertainment satellite. The audio tuning optimizes the sound for each channel and also allows for the tuning of other audio entertainment programs found on some channels. The power switch activates the receiver from the remote control location. A red light-emitting diode (LED) indicates receiver operation.

The control console and the receiver unit are linked by a 10-conductor control cable. The receiver unit is housed in a weather-resistant case and is mounted as close to the dish antenna as is practical. Common, low-cost RG-59 cable carries the satellite signals from the receiver to your television. The receiver converts the microwave signals from the LNA to a 70 MHz signal, which is then demodulated to video and audio signals usable by a TV monitor. A TV monitor is not the same as your standard television receiver. For the latter, reception will be had with the addition of KLM's KM-2 Modulator, which is an option.

Fig. 12-10. Open chassis view of Sky Eye II kit (courtesy KLM Electronics, Inc.).

The receiver has no controls and requires no adjustments after build-ing and installation. All cables except the ac power cord on the control console carry only low voltages. While this is a kit, the microwave conver-ter and tuning circuits have been factory-assembled, tested, and aligned. KLM states that no microwave test equipment is needed for the assembly and alignment of the KLM Sky Eye II receiver kit. As was mentioned earlier in this text, the assembly and alignment of microwave frequency-determining elements is really no job for the home experimenter. KLM has wired the hard parts but leaves a great deal of the more standard wiring and alignment for the builder.

Figure 12-10 shows an open-chassis view of the Sky Eye II kit. The master receiver unit is on the left, the remote control assembly on the right, and the five circuit boards that encompass the kit are in the foreground. This is identical to the completely wired version offered by KLM as far as circuitry and performance characteristics are concerned. Figure 12-11 gives the technical specifications for the assembled kit.

The KLM Sky Eye II satellite receiver kit is not intended for first-time kit builders or the electronics novice. You will need some previous kit-building experience plus some knowledge of radio or TV alignment. If this is your first kit or you are relatively unfamiliar with rf or video electronics, you will need the help of an experienced electronic assembler or technician.

Assembling the Sky Eye II kit can be an enjoyable and rewarding experience if you follow a few simple guidelines. Pick a quiet, well-lit area for assembly. Don't work on the kit when you are overly tired or preoc-

FREQUENCY RANGE:	3.7-4.2 GHz
TUNING BANDWIDTH:	500 MHz
NOISE FIGURE:	13 dB
INPUT IMPEDANCE:	50 ohm, nominal, unbalanced
INPUT CONNECTOR:	Type "N" female
CHANNELS:	24 nominal
OUTPUT VIDEO BASEBAND:	4.5 MHz
OUTPUT AUDIO BANDWIDTH:	to 10 KHz
OUTPUT SIGNAL LEVEL:	1V peak to peak
OUTPUT IMPEDANCE (VIDEO):	75 ohms, nominal, unbalanced
OUTPUT CONNECTOR (VIDEO):	Type "F" female
OUTPUT CONNECTOR (AUDIO):	Type "F" female
OUTPUT CONNECTOR (RF) (optional)	Type "F" female
INPUT POWER:	115 VAC nominal, 50-60 Hz,
DIMENSIONS:	
RCVR	12.5″ × 12.5″ × 8.5″H
Control Console	6″W × 7″L × 3.5″H

Fig. 12-11. Sky Eye II receiver specifications (courtesy KLM Electronics, Inc.).

cupied with other matters. Follow the instructions in each section of their assembly manual carefully and in the sequence presented. Review each section before starting and read the entire step before performing it. If possible, give yourself enough time to complete an entire page at one sitting. Return unused parts to the hardware bag supplied with the kit when stopping. Also, leave the assembly area neat, with some indication as to where assembly is to be resumed.

Circuit board illustrations are 120% of actual size for clarity. Parts are drawn in black. Shaded areas usually indicate foil areas, although this scheme is sometimes reversed for clarity. Solder points are indicated by black dots, circles, or bars. Component numbers in the illustrations match the numbers in the instruction boxes. An asterisk in an illustration indicates attention to a polarity factor for a particular part. Accompanying sketches are included to assist you in locating and installing certain parts or assemblies. Caution is necessary in the assembly process when drilling or clipping parts leads, as particles can fly into your eyes. Safety glasses or goggles are recommended.

I was most impressed with the instruction manual that is included with all Sky Eye II receiver kits. Figure 12-12 shows a sample of the main chassis wiring instructions. Note that large line drawings are used to illustrate each step-by-step procedure. The builder simply reads the instructions, performs the operations outlined, and then places a checkmark to indicate that this phase of assembly has been completed. The fact that the drawings are 120% if the actual circuit board size is a big help in clarifying all assembly procedures.

Figure 12-13 shows a block diagram of the Sky Eye II receiver. The amplified signal from the antenna and LNA is routed to the mixer module. The signal is mixed with the output of a voltage-controlled oscillator to produce a difference frequency of 70 MHz. The 70 MHz i-f signal is amplified by a pair of limiting amplifiers (first and second i-f amplifier modules) and routed to the video demodulator.

The video demodulator contains a FM discriminator and video shaping amplifiers. The output of the video demodulator contains the TV video signal and the FM audio subcarrier. A separate output of the video demodulator is routed to the audio demodulator module. The audio demodulator

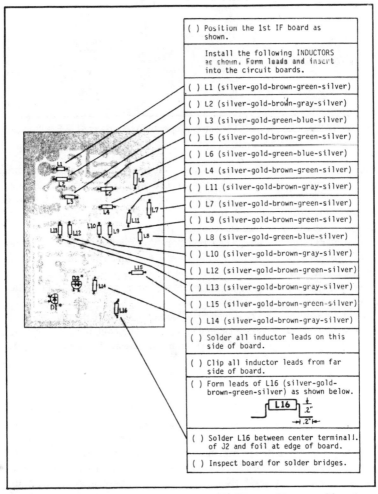

Fig. 12-12. Typical instruction page from Sky Eye II assembly manual (courtesy KLM Electronics, Inc.).

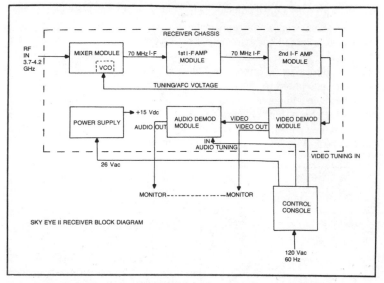

Fig. 12-13. Block diagram of KLM receiver operation (courtesy KLM Electronics, Inc.).

contains a 10.7 MHz FM receiver and a voltage-controlled oscillator. The FM audio subcarrier is demodulated to recover the TV audio signal.

The video and audio from the demodulator modules is amplified to a voltage level of one volt peak-to-peak and routed to the TV monitor, video recorder, or TV modulator via coaxial cables.

Power is supplied by a bridge rectifier, filter, and solid-state regulator power supply located on the receiver unit. A transformer in the control center provides 26 Vac for receiver operation. Tuning voltages for the audio and channel select functions are provided by potentiometers in the control center.

The receiver unit is designed to be located as close to the antenna feed horn as possible for improved signal reception. The control center may be located at the viewing location (100 feet or more away) for tuning convenience.

MIXER MODULE

Referring to Fig. 12-14, the 3.7-4.2 GHz frequency band from the system LNA is routed to mixer input connector J1. The mixer module is assembled at the factory. A factory-supplied coaxial transition section is used between the antenna transmission coaxial cable and connector J1. Regulated voltage for the LNA is connected to the center conductor of the coax cable via C2 and RFC1. (Note: The LNA voltage connection is used for KLM-supplied LNAs. Other LNAs may require a different arrangement and the +15 Vdc may be disconnected from the feedthru capacitor.)

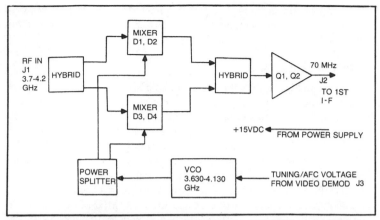

Fig. 12-14. KLM receiver mixer schematic (courtesy KLM Electronics, Inc.).

The input signal is routed to a pair of mixers D1-D2, D3-D4 through stripline hybrids. The input signal is phased shifted by 90° to each mixer. The local oscillator frequency in the band 3.630-4.130 GHz is supplied by the voltage-controlled oscillator (VCO). The pair of mixers and the hybrids comprise an image rejection mixer. The output frequency is the difference frequency of 70 MHz and is routed to a non-limiting amplifier consisting of Q1, Q2, and associated components. The tuning voltage for the vco is obtained from the automatic frequency control (afc) circuits of the video demodulator and routed to the mixer module via coaxial connector J3.

FIRST I-F AMPLIFIER

Figure 12-15 shows a schematic of the first i-f amplifier/limiter. The 70 MHz i-f signal from the mixer module is received at input connector J1 and routed through a bandpass filter consisting of L1-L8, C1-C7. Variable capacitor C1 and L1 are tuned to a nominal 85 MHz. Variable capacitor C6 and L6 are tuned to a nominal 55 MHz. Hybrid amplifiers U1, U2 and U3 form a three-stage progressive limiting amplifier. The filtered and

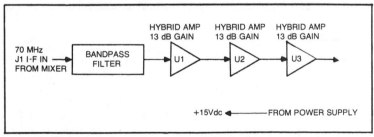

Fig. 12-15. Schematic diagram of first i-f amplifier/limiter (courtesy KLM Electronics, Inc.).

235

amplified 70 MHz signal is routed to a connector J2 and to the second i-f amplifier module. Regulated +15 Vdc power is supplied via feedthru capacitor C22.

SECOND I-F AMPLIFIER

Shown in Fig. 12-16 is the second i-f amplifier/limiter schematic. The 70 MHz i-f signal from the first i-f amplifier is received at input connector J1 and routed through a bandpass filter consisting of L1-L8, C1-C8. Variable capacitor C1 and inductor L1 are tuned to a nominal 85 MHz. Variable capacitor C6 and L6 are tuned to a nominal 55 MHz. Hybrid amplfiers U1 and U2 and associated components from a two-stage progressive limiting amplifier. The gain of the first i-f amplifier and the second i-f amplifier is such that the output stage U2 will begin to limit when the input to the first i-f amplifier is at −30 dBm. The filtered and limited 70 MHz i-f signal is routed to connector J2 and to the video demodulator module. Regulated +15 Vdc power is supplied via feedthru capacitor C19.

VIDEO DEMODULATOR

Figure 12-17 provides a block diagram of the video demodulator. The 70 MHz i-f signal from the second i-f amplifier is routed to input connector J1 and to the input of amplifier Q1. Variable capacitor C3 and inductor RFC1 are tuned to a nominal 90 MHz. Variable capacitor C5 and inductor RFC2 are tuned to a nominal 50 MHz. Transistors Q1-Q3 and diodes D1-D4 and associated components form a 70 MHz FM discriminator. Variable capacitor C14 and inductor RFC3 are tuned to a nominal 82.5 MHz. Variable capacitor C19 and inductor RFC4 are tuned to a nominal 57.5 MHz. C14 and C19 form the discriminator linearity adjustments. The discriminated video signal from the output of D1-D4 is routed to video amplifier Q4 and to afc operational amplifier U1. The video signal from the emitter of Q4 is routed to connector J2 and to the audio demodulator module. The video signal from the emitter of Q4 is also routed through a de-emphasis network consisting of R25-R29, C28-C31, and RFC5-RFC6 to the input of amplifier U2. Amplifier U2 and transistors Q4, Q6 and associated components form the video output driver amplifier. The amplified video output is at a level of one

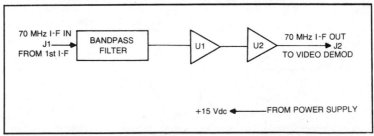

Fig. 12-16. Second i-f amplifier/limiter schematic (courtesy KLM Electronics, Inc.).

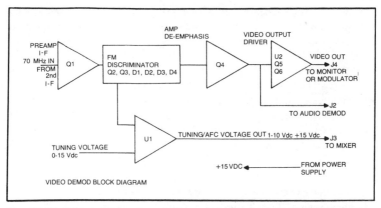

Fig. 12-17. Schematic drawing of video modulator (courtesy KLM Electronics, Inc.).

volt peak-to-peak and is routed to the external video monitor, recorder, or modulator via connector J4. Diode D7 and associated components are a clamping network to eliminate the 30 Hz video framing signal.

Tuning voltage from the remote control center unit is routed to the afc amplifier U1 via feedthru capacitor C39. The remote tuning voltage is used as a steering voltage for the afc amplifier. The voltage from the discriminator is applied to the other input of the afc amplifier and the resulting afc voltage is used to control the vco in the mixer module. The afc feature ensures that the receiver will remain tuned to the desired channel. The tuning/afc voltage is routed to the mixer module via connector J3.

AUDIO DEMODULATOR

The audio demodulator block diagram is shown in Fig. 12-18. The audio demodulator module is a tunable FM receiver which is used to recover the TV program audio signal or a background music channel. The video signal from the video demodulator module is routed via a network consisting of C28, R1-R4, R7, RFC1, and C1-C2 to mixer U1. The video signal is mixed with a local oscillator carrier in the range 15.7 to 18.7 MHz to produce an output signal at 10.7 MHz. The 15.7 to 18.7 MHz tunable signal is generated by voltage-controlled oscillator U2. The tuning range is adjusted to center the FM audio subcarrier at 10.7 MHz at the mixer output. The output signal from mixer U1 is routed through ceramic resonator bandpass filter Y1 to the input of U3. Integrated circuit U3 is a complete FM receiver i-f, discriminator, and audio section. The discriminator tuned circuit is comprised of ceramic resonator Y2 and associated components. The audio output of U3 is amplified by an amplifier consisting of one half of dual amplifier U4 and routed to the external TV monitor, video recorder, or TV modulator via connector J2. The audio output signal level is a nominal one volt peak-to-peak.

Vco U2 is tuned over the frequency range 15.7 to 18.7 MHz by a tuning voltage from the control center via feedthru capacitor C25. The tuning voltage is routed to one input of an afc amplifier consisting of one half of dual amplifier U4. Afc voltage from discriminator Y2 is routed to the other input of the afc amplifier. The output of the afc amplifier is the tuning/afc voltage which controls the output frequency of the vco. The afc feature ensures that the audio demodulator will stay tuned to the desired audio subcarrier.

Solid-state regulator U5 is used to provide regulated +6 Vdc to the vco. The +15 Vdc power to the module is received from the power supply via feedthru capacitor C26.

POWER SUPPLY

Referring to Fig. 12-13, a transformer located in the control center is used to provide 26 Vac via the remote cable to the receiver chassis. The ac voltage is rectified by bridge rectifier CR1 and filtered by capacitor C1. The filtered voltage is routed to a solid-state regulator U1 and regulated to +15 Vdc. Ceramic capacitor C2 is a bypass for high frequency noise at the regulator. Filtered dc at approximately 36 V is routed to the control center via the remote cable.

CONTROL CONSOLE

Input 115 Vac power is controlled by a power switch. The 26 Vac output of the transformer is routed to the receiver unit power supply via a connector and the remote cable. Filtered +35 Vdc from the remote power supply is returned to the control console and routed to an LED and 26 Vdc regulator. Regulated +26 Vdc is routed to a channel tuning potentiometer and variable resistor. Tuning voltages from the resistors are routed to the video demodulator and audio demodulator modules in the receiver unit via a connector and remote cable.

INSTALLATION

The installation of the Sky Eye II receiver is dependent upon the location of your high gain antenna and your TV monitor or recorder. A few basic guidelines are presented to aid you in this procedure.

The antenna should be located so that an unobstructed line-of-sight view of the desired satellite or satellites is obtained. If possible, the antenna should be located reasonably close to the viewing location. Space available, aesthetics, zoning ordinances, neighborhood wishes, as well as operating requirements, may dictate the final location of the antenna.

The receiver unit should be mounted on the antenna mount or as close to the antenna as possible. Total length of RG-142 cable from the LNA to the receiver must not exceed 25 dB attenuation (about 100 feet). The aluminum channel (heatsink) on the bottom of the receiver must be used as a mounting surface. The channel may be attached to the antenna mount, a post or mast, or other attaching device. Bolts, U-bolts, lag screws, etc.,

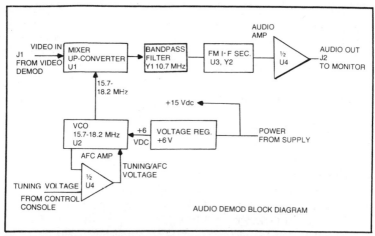

Fig. 12-18. Audio demodulator diagram (courtesy KLM Electronics, Inc.).

may be used for attachment. The channel may be drilled as necessary as long as care is taken not to damage the power supply voltage regulator. The receiver should be placed so that the black weather cover is horizontal. Additional covering such as plastic sheeting, a ventilated box, or the eave of an adjacent building may be employed if severe weather conditions are expected.

The control center should be placed at or near the TV monitor, video recorder, or TV receiver. The control center may be up to 100 feet from the receiver unit. A longer distance is possible if the internal control center tuning voltages are readjusted. The control center power cable may be plugged into any three-wire grounded household receptacle which supplies 115 Vac at 50-60 Hz.

SOLDERING

As with any electronic kit, soldering is the most important operation performed. A bad solder connection can easily prevent your satellite receiver from working properly. This condition can even damage some internal components. A good solder connection forms a permanent electrical contact between two parts and allows the assembly to function as it was designed.

As outlined in the KLM assembly manual, it takes only a few simple rules to make a good solder connection each time:

■ Use a small (25 to 40 watt) soldering pencil with a chisel tip (preferred) or a pyramid tip.

■ Keep your soldering iron tip clean at all times. Wipe the tip on a damp cloth or sponge and then immediately reapply fresh solder for a wet and shiny appearance. Use only quality resin core solder, 60/40, which is fine enough to be applied in cramped areas. Under no circumstances should

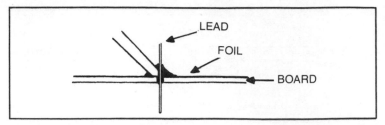

Fig. 12-19. Correct soldering procedure.

acid core solder be used (this is designed for plumbing work and can completely ruin electronic circuits).

■ A good solder connection is formed when the tip of the soldering iron is placed in a position that will heat the component lead and the circuit board foil at the same time. This is shown in Fig. 12-19.

■ A poor solder connection is formed when the tip of the iron is only contacting the foil of board. This is shown in Fig. 12-20. Here, not enough heat is transferred to the component lead, and the resin forms a coating to insulate the lead from the foil.

■ A poor solder connection may also be formed when the tip of the iron is only contacting the component lead, as shown in Fig. 12-21. In this situation, an insufficient amount of heat is transferred to the foil which is insulated by the resin.

■ Be alert to solder bridges, particularly when soldering in cramped spaces. A solder bridge forms when another foil trace or lead is touched by the iron when soldering. This usually happens when too much solder is used or the iron is allowed to drag across other circuit leads. When soldering the circuit board in the KLM kit, take note of adjacent foil areas and place the iron for best foil/lead contact only with the part you wish to solder. Use only enough solder to make the connection. To remove a solder bridge, turn the board over so the excess solder will flow down to the iron when heat is applied.

■ A properly made solder joint will have a shiny appearance. Improper connections will be dull and may form in small blobs around the component lead and foil.

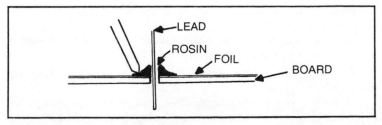

Fig. 12-20. A poor solder joint may be formed when the iron is placed only on the circuit board foil.

240

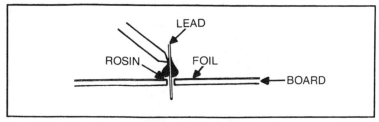

Fig. 12-21. When the soldering iron is placed only on the component lead, bonding may not take place at the proper foil.

These soldering techniques apply to almost any electronics kit and not to only the KLM satellite receiver.

RECEIVER ALIGNMENT

The KLM Sky Eye II receiver can be aligned to obtain a usable picture with only a TV monitor and voltmeter. However, other instruments should be used in this procedure to assure top notch performance. These will be discussed later in this chapter. Before beginning, complete the assembly of the remote control cable and the video and audio cables as detailed in the assembly manual. The receiver is now connected to the control center unit with remote cable.

Step-by-step test and alignment procedures for the power supply, control center, and receiver module are contained in the instruction manual. For the power supply, the ac power cord is connected to a 115 Vac 50-60 cycle source and the 10 conductor cable is attached between the control console and the receiver unit. From here on, the process involves running checks of voltage values at the pins of the receiver unit connector. The proper readings are provided in an appropriate table included with the alignment instructions. Alignment of the i-f amplifiers is accomplished in eight steps for each. A sweep generator, oscilloscope, and 75-ohm detector are required for maximum performance. These units are set up as shown in Fig. 12-22.

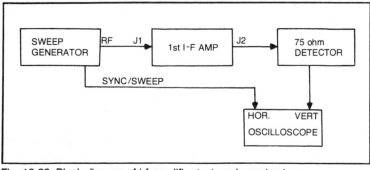

Fig. 12-22. Block diagram of i-f amplifier test equipment setup.

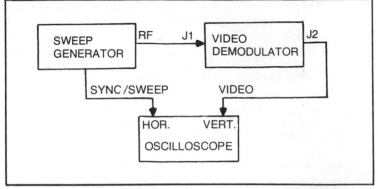

Fig. 12-23. Alignment of video demodulator requires a sweep generator and oscilloscope.

Alignment of the video demodulator requires the sweep generator and oscilloscope again, which are set up as shown in Fig. 12-23. This stage is aligned in eleven steps.

Other adjustments include setting the output gain on the video demodulator, aligning the control center, and a few other steps which are easily and quickly accomplished.

Then end result is a satellite television receiver which should perform at least as well as the commercially wired unit. Naturally, this assumes that the builder follows all steps as outlined by KLM and exercises great care in soldering and all other building procedures. Again, due to the nature of this kit, it is definitely not recommended as a first-time project. Before attempting to build it, you should have good experience in assembling other types of electronic kits that operate at lower frequencies. If you have assembled several Heathkit devices, for example, you will most likely be very qualified to tackle the KLM satellite receiver kit. Actually, the construction techniques involved are no more difficult than those incurred when putting toegether an amateur radio transceiver. When you finish your project, you can take pride in the fact that you have built most of it yourself. It works perfectly and you saved nearly $800.

SUMMARY

While there are relatively few home assembly kits presently available to the TVRO Earth station enthusiast, the few that are offered present experimental possibilities along with some that can be directly incorporated into an Earth station when completed. As of this writing, several other companies have expressed their intentions of offering satellite receiver kits which are carbon copies of the commercial models they presently offer in completed form.

As the TVRO industry grows, there will certainly be more and more opportunities for the individual who likes to build his own equipment to obtain kits which can be efficiently assembled at home. For now, the frequency-determining microwave circuits will probably be left up to the manufacturer who must assemble them and make certain that they are operational before being delivered as components in electronic kits. As microwave technology advances, it is conceivable that discrete microwave assemblies will be made available. These will allow the home builder to design a receiver more or less from the ground up. Here, the builder would purchase all of the more standard parts as discrete components and design them around the microwave circuits which he buys in completed form.

I am a firm believer in constructing as many devices and circuits in a home shop as is possible. The builder gains a better understanding of how the circuit operates and is in a much better position to perform repairs should they be needed. Home construction also leaves room for serious experimentation which can ultimately result in improved circuit performance. Most of these advantages cannot be had with a commercially manufactured and assembled device.

Editor's Note: Basic information on home-brewing parabolic reflectors is contained in TAB Book No. 1367, *The GIANT Book of Electronics Projects* by the Editors of 73 Magazine.

Chapter 13
Specific Site Selection

As is always the case when setting up any antenna system, the physical site is of utmost importance. Previous chapters in this book have outlined methods of choosing a general site, one which has windows open to the satellites you wish to receive. This selection was done with the aid of a computer printout for a general geographic area. This lets you know whether or not it is possible to receive certain satellites, but it does not mean necessarily that adequate reception will be maintained at a specific site within this general area. The computer printout does not take into account smaller mountains, metal towers, large trees, and other obstructions which can severely interfere with the satellite transmission path. This must be taken into account when you choose a specific mounting site on your property, because one location may be poor for certain satellites, while another a few feet away may be ideal.

First of all you will want to obtain the computer printout, then with a compass and inclinometer, determine the exact position in the sky toward which your antenna must be aimed for reception of a particular satellite. Building a simple sighting device may help you. This can be constructed from a small piece of lumber or dowel rod, which is fitted with two nails. This is shown in Fig. 13-1. By sighting along the nail heads, as shown in Fig. 13-2, you will be in a good position to see whether or not you have any major obstructions.

If you live in an urban area, buildings and metal towers often abound and can present major interference problems. But moving the dish site a few feet in one direction or another will often remedy this problem, at least for a particular satellite. Fortunately, many urban apartment buildings are constructed with flat roofs which can make ideal mounting sites for

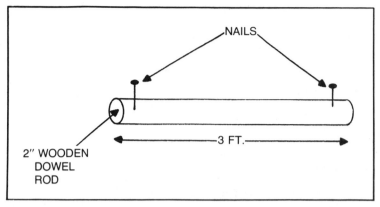

Fig. 13-1. A simple sighting device can be constructed from a piece of lumber fitted with two nails.

parabolic dishes. When located atop a high building, many terrestrial obstructions are cleared and many more clear windows to various satellites should be available.

The antenna's view through the window can be thought of as being the size of the dish diameter. For this reason, you will want to make certain when sighting on a satellite with your homemade wooden device that the window is open to all portions of the antenna. The best way to do this is to place the sighting device at the exact center of the mounting site. Once you have obtained a clear window, move the site one-half the diameter of the antenna to the right and left of a line which is perpendicular to the satellite

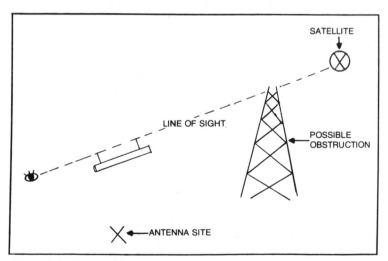

Fig. 13-2. By sighting along the nail heads you can determine if there are any obstructions between your site and the satellite.

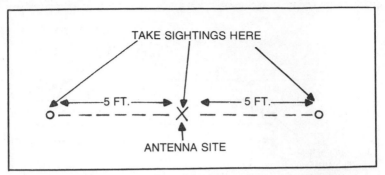

Fig. 13-3. To insure that all portions of the antenna have a clear window, move the site five feet to the right of the center of the dish antenna.

signal path. For a ten-foot dish antenna, you would move the site five feet to the left of center and then five feet to the right of the same point, taking sightings at each additional location. This is shown in Fig. 13-3. Be sure to keep the site at the same elevation and azimuth for each reading. This will assure a completely open window for the antenna to intercept the transmissions of a particular orbiter. Repeat these steps for each additional satellite you wish to receive. It may take many small repositionings of the antenna's center mounting site before the best location is determined.

Small trees and bushes are normally not considered to be major signal obstructions, although in marginal areas these should be avoided whenever possible. Any object or structure which is made of metal, especially those that are very large, can play havoc with reception, so these are to be avoided in all systems regardless of relative signal strength at your particular geographic location. It may be necessary to locate the antenna at a higher physical site to overcome these, although in some instances a site which is lower in elevation may be able to look under these potential sources of interference. Think of the sighting routine as firing a gun at a target. In other words, you want to hit the satellite with the fired projectile. If there are no major objects in the way, your bullet will hit its mark. Any obstructions lessen the chances of a successful hit.

Hills and mountains are often encountered in rural areas and present major interference problems. This is especially true if your antenna is mounted at the base of one of these, as shown in Fig. 13-4. The only solution is changing sites if satellites from behind the mountain are to be received. You might move the dish to the top of the mountain or to a point farther away from the base, as is shown in Fig. 13-5. Make sure there is plenty of clearance between the mountain ridge and the antenna sighting line to assure maximum signal reception. When site relocation is not possible, you will simply have to be content or receive the signals of those satellites which lie out of the shadow of the hill or mountain.

Once you have obtained a good potential site location, you will want to examine some of the other physical characteristics which may prove to be

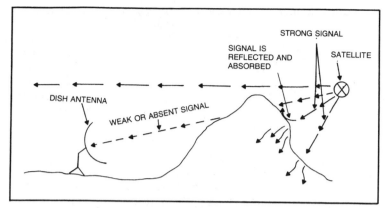

Fig. 13-4. You may have reception problems if your antenna site is located at the base of a mountain or hill

very important in the future. There are very few demands made upon a TVRO Earth station site when compared with some other types of antennas used for operation at a lower frequency. Some types of antennas must be mounted over soil which exhibits good electrical conductivity. This is not even a consideration with TVRO installations. The parabolic dish does not depend upon soil conductivity to operate properly.

Putting aside the unimportant electrical characteristics, the physical condition of the site will involve the firmness of the soil, the presence of large underground rocks, accessibility, etc. When installing most dish antennas, it is necessary to bore three or more holes several feet into the earth. These will later be filled with concrete which will serve as anchors for the antenna mount. Obviously, if you have very rocky soil conditions, it's

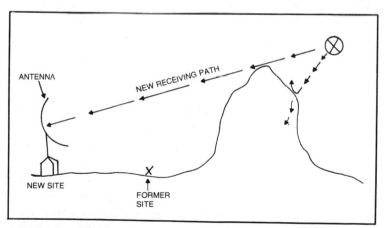

Fig. 13-5. It may be necessary to relocate your antenna site to a point farther away from the base of a mountain.

going to be difficult to drill these holes on your own. Hiring a commercial crew to come in and do it for you can be rather expensive.

TVRO Earth station enthusiasts should be very careful when the dish antenna must be mounted in swampy areas. The high water content of the soil can cause the concrete footers to shift during changing weather conditions. If the base shifts even slightly, the antenna will be out of alignment with the satellite and reception will be greatly reduced or possibly terminated completely. Certainly, the antenna can be readjusted, but this may have to be done quite often as soil conditions change. Swampy areas may be successfully used, but it is often necessary to sink the concrete footers much deeper. This adds to the site preparation expense.

Along these same lines, you will want to check land contours around the proposed site location. If you choose a low spot which serves as a natural drainage and collection point for runoff water, the same problem mentioned above can exist.

How secure is your proposed site? Is it off the beaten path? Or is it near an area which is quite accessible to passersby, who may want to vandalize an attractive setup like a TVRO Earth station? If an easily accessible site is chosen, it would be wise to go to the extra expense of constructing a fence or other type of barrier around the installation to prevent unauthorized access. Standard practice is to place a few "Danger! High Voltage" signs on the antenna and Earth station. This may be of some help, but is often an exercise in futility if a physical barricade is not used as well.

Another site selection item which must be considered is that of trees and other tall objects which are located nearby. While these may not interfere with the satellite transmission window, in the spring a leafy growth can expand the size of a tree by a fair amount. Appropriate trimming during the growing season will provide an easy correction of this problem. Another major factor to be considered about the earth site surroundings is whether or not large trees could fall on the antenna and equipment should a heavy wind bring them down. One sizeable tree tearing through your Earth station can instantaneously reduce the entire system to rubble. An Earth station is an expensive undertaking and must be protected at all times from all dangers.

In growing residential neighborhoods and in commercial districts, it might be a good idea to run a check at your local city hall to see if any building permits have been issued for nearby structures which could become potential sources of interference. That empty lot next door may present no problem now, but when a house or other building is erected there, it could totally ruin your perfect window. Once an Earth station is installed, it can be thought of as semi-permanent. While the location can be moved, this will require the drilling of more holes for establishment of the new base and, of course, the labor involved in the dismantling and reassembling of the station. While you're at City Hall, also check on the zoning regulations to see if any local ordinances prohibit the erection of an Earth station antenna. In some areas, permits may be required. In others, you may have to appear before a Board of Zoning Appeals in order to obtain a variance to a restrictive ordinance. These are not hard to get, but you have to stress the fact that there is no high voltage danger, nor any chance of the structure damaging property should it fall from its mount. Most antenna ordinances involve height limitations which most Earth-mounted TVRO antennas do not even come close to exceeding.

The problems mentioned here and some possible solutions are generalized in that they could apply to any part of the country. By regionalizing this discussion, a few basic facts can be presented. Since the satellites we are interested in are in geostationary orbits over the Equator, those persons living closer to the Equator will generally have fewer problems with earth-mounted objects, since the antenna elevation angle will be increased. In other words, the antenna will be pointed higher in the sky. Antenna sites in the northern part of this hemisphere will have to be aimed lower on the horizon. This can bring distant mountains and other natural objects into the path of the satellite transmission window. This is illustrated in Fig. 13-6. Of course, when an antenna is pointed higher in the sky, this can result in other problems. This does not apply so much to parabolic TVRO antennas as it does to spherical designs discussed in ealier chapters. The spherical dish is not adjusted; rather, the feed horn is moved for reception of different satellites. If the spherical dish is aimed toward a high portion of the sky, this will mean that the feed horn will have to be mounted a higher distance from the ground than when the dish is lowered to a point closer to the horizon. This is shown in Fig. 13-7.

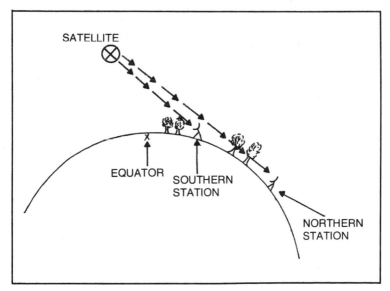

Fig. 13-6. Antenna sites in the northern part of this hemisphere will be aimed at a lower point on the horizon.

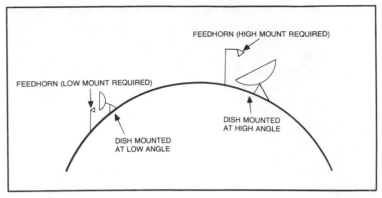

FEEDHORN (HIGH MOUNT REQUIRED)

FEEDHORN (LOW MOUNT REQUIRED)

DISH MOUNTED
AT HIGH ANGLE

DISH MOUNTED
AT LOW ANGLE

Fig. 13-7. When a spherical dish is pointed higher toward the sky, it will be necessary for the feed horn to be mounted higher from the ground.

In rural locations, site access is sometimes a problem, Remember, you will have to truck the heavy components to the site, and while the finished Earth station consumes very little area, several times this amount of area will be needed as a work site to put the whole thing together. A small porch roof may make an excellent mounting site for an Earth station, but it will be difficult and potentially dangerous to assemble one there, since all of the parts will have to be spread out and assembled piece by piece. Additionally, two or three men will be required to assemble the antenna panel sections. Small space problems can be overcome during the assembly, but the extra preparations and precautions must be planned in advance. Remember also that you will probably have to travel to the site every so often for preventive maintenance and to adjust the antenna to receive a different satellite. During heavy snowstorms, dish antennas must be cleaned periodically to avoid excessive ice and snow buildup which can cause structural damage as well as interfere with reception. If you find it necessary to locate the antenna some distance from the house and use line amplifiers to boost the signal strength caused by the increased transmission line lengths, make certain you can get to the site in all types of weather. Even if you have motorized antenna positioning controls which are activated at the viewing location, regular on-site inspections are still necessary.

Another consideration is electric power. If the antenna is mounted a short distance from the home, this can probably be handled by means of a waterproof extension cord. But if larger distances are involved, you may wish to run an underground service to the antenna site. This is especially true when motorized base drives are to be used. In some areas, a separate electrical installation may require a permit and/or an electrical inspection. This is to make certain that your installation conforms to all safety regulations.

Wind problems have been discussed many times in this book and it bears repeating again under this site selection discussion. Most commercially manufactured Earth station antennas are designed to withstand 100 mile per hour winds when mounted on approved bases and in compliance with the manufacturers' instructions. This does not necessarily mean, however, that the antenna will efficiently receive signals under these conditions. As structurally sound as most Earth stations are, high winds can cause the dish antenna to vibrate and play havoc with the received signal. If the site you have selected is regularly exposed to high gusting wind velocities, you may want to consider building a wind break to one side of the dish directly in the path of prevailing winds. A local contractor may be able to help you with the design and construction. In no instances should the wind break be allowed to interfere with the sighting path of the antenna. Figure 13-8 shows how a protective structure might be installed in relationship to the Earth station. This structure is not absolutely essential but will provide more reliable reception characteristics from your Earth station during high wind conditions.

The problems and corrective measures discussed in this chapter assume that the TVRO Earth station owner has some flexibility in where the site is to be located. From a practical standpoint, this is not true in many cases. Most of us will be limited to a site in a backyard or a short distance from the home. In these situations, you do the best you can. In some areas, it simply will not be practical or possible to efficiently receive the signals from all of the satellites which have open windows to your geographical area. Here, you will have to choose the location within the area allotted to you which provides the best reception of the largest number of potential satellites. Compromises and tradeoffs will certainly be mandated here. Of course, it may be possible to get a neighbor involved who might allow you to use a small portion of his property if this location provides a better mounting site. A small rental fee could be paid each month for this privilege, but more often, your Earth station's landlord will want a free tap-in. A signal

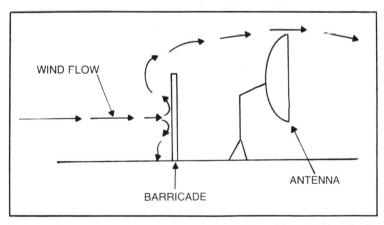

Fig. 13-8. A protective structure such as this would provide a wind breaker in areas where winds tend to be severe.

splitter would be required and might be the cheapest way to go when the total cost is spread out over a few years of operation.

Some homeowners find that they must locate their stations in the front yard rather than in the relatively hidden recesses of the backyard. If you must do this, be certain to provide some means of barricading the station from prying eyes and hands. A sturdy chain link fence with barbed wire at the top is rather costly but very effective, although it does little for the appearance of your home. If your entire front yard is fenced in, a locked gate may serve almost as well. Some owners who find themselves in a situation which requires a front yard installation buy large tarpaulins which are used to cover the antenna and other components during times when no one is home. When vacations are planned, it's a good idea to have a neighbor keep an eye on things for you. It is also wise to partially disassemble the station by removing the LNA and any antenna-mounted electronics. This effectively leaves only the dish accessible. It's pretty hard to steal one of these without being seen, and the antenna is the most rugged portion of the station, so it is not as easily damaged by vandals.

Further protection of an Earth station can be provided by installing some sort of an alarm system. These can be built from parts obtained from your local hobby store or installed by professionals. Proximity alarms are triggered whenever anything enters the protected area. No physical contact with any portion of the alarm is necessary. These types of systems have excellent applications when a physical barrier is placed around the TVRO Earth station, but when the barrier is not present, dogs, cats, and other stray animals can keep the alarm sounding all night. Many TVRO Earth station receivers are now equipped with warning circuits which will trigger an external alarm whenever the low noise amplifier is disconnected. In order to steal an LNA, it is necessary to disconnect its power supply and cable. This will automatically trigger the receiver alarm in those units which are so equipped. The low noise amplifier and the downconverter which often is attached represent a great deal of expense and could be potentially high-priority rip-off items for persons who are so inclined.

Of course, the best way to prevent theft and vandalism is to hide the Earth station as effectively as possible. This is not easily done, especially with that large dish antenna sticking out into space for all the world to see. For this reason, a backyard location is always preferable.

For some readers, this discussion will be mostly academic, as most persons don't select a site so much as it selects them. Fortunately, there are enough orbiting satellites, each carrying a significant number of program services, to make it worthwhile to install a station which receives only one or two. Most owners home in on one prime satellite for 95% of all viewing. The other 5% of the time, the system is used for experimental purposes to see just how many satellites can be received. Of course, more and more satellites are planned for the near future, so it will behoove the Earth station owner to choose as flexible a site as is possible.

The selection of an appropriate TVRO Earth station site will be the prime criteria for all future reception uses. A poor site choice will result in poor selection and flexibility, regardless of the quality of the Earth station components. It is for this reason that all potential Earth station owners should spend the hours required in researching this basic aspect of satellite TV reception well before any equipment is purchased and any construction is attempted.

A very few individuals will find that they are just not in a position to install an Earth station which will provide satisfactory performance. There is no one quite so miserable as the person who spends many thousands of dollars for a piece of equipment which *cannot* do what he wants it to do because of location. This will be the rare exception rather than the rule. If you can receive only one or two satellites from your location (and receive them well) then the work and expense involved in owning your own TVRO Earth station will be worthwhile. If you are contemplating a move in the near future and your available site is not really appropriate, I suggest that you wait awhile. Maybe your new location will provide several prime sites for the installation of an efficient receiving system.

Don't make a decision until you know all the facts. You can start your research by purchasing a computer printout of windows in your area. Then, using the homemade sight and a few other inexpensive instruments, you can determine very accurately exactly what your location has to offer. Too many persons have become so avid in their quest for a TVRO Earth station that they have neglected this prerequisite. Most have gotten away with it very handily, but a few have been burned. Don't be one of the latter. The thirty or forty dollars you spend for the printout and the simple instruments, added to the couple of hours time spent in recording measurements, will pay off very handsomely for you when the time comes to purchase your personal TVRO Earth station equipment.

Chapter 14
The Earth
Station Assembly

The purpose of this book has been to provide the reader with the information necessary to make intelligent decisions regarding what equipment to consider and ultimately, to buy for assembling a complete TVRO Earth station. This last chapter will serve as a summary of all of the information contained in this text and to describe what the assembly of a complete TVRO Earth station will involve. The appendices are filled with charts and other materials which will aid you in the initial planning stages and which may be used in the years to come as a convenient reference source.

SITE DETERMINATION

Your first duty as one who is interested in installing a personal TVRO Earth station is to select a site. This will involve obtaining a computer printout for your geographical area to determine the number of satellites which you have a reasonable chance of accessing from your area. When you have this information, you will most likely be in an excellent position to see how much use you will be able to benefit from an Earth station.

If the computer scan looks favorable for your area, you will then select a specific site near your home at which the antenna and other TVRO equipment is to be mounted. Many criteria for selecting a favorable site have been provided in an earlier chapter, and you may want to refer to this when choosing between several possible TVRO locations.

As is outlined, you will want to run an "open window" check, which will allow you to detect any interfering objects which may be situated between the proposed antenna location and the satellites you wish to receive. This check will provide the necessary information as to which satellites generally available to your area may be specifically received by your station.

During this entire process, you will also want to refer to the relative signal strength charts provided in the appendices of this book to determine how strong the satellite transmissions are in your area. Chances are, the manufacturers of specific Earth station antennas and other components will also supply you with the same basic charts. The latter are often coded with specific antennas and LNAs recommended by these manufacturers for stations in different areas of the United States.

Having done all this, it is now necessary to determine the physical nature of your site. Referring to the previous chapter on this subject, we know that the site has to be accessible and fairly stable in regard to soil conditions in order to provide a firm base for the antenna. From the information contained in this section of the book, you will probably be able to determine whether or not additional site preparation will be necessary before beginning construction of the antenna mounting base.

EQUIPMENT SELECTION

After you have determined that you are indeed in a situation to make full use of a TVRO Earth station, the fun part begins. You will have to decide on which manufacturer or manufacturers you will select to supply the TVRO Earth station components. You may even elect to build some components yourself. In any event if you are like most TVRO enthusiasts in the country, you will probably purchase all of your equipment from one manufacturer. Many products from a variety of companies which sell TVRO Earth station equipment have been mentioned in this book. Additionally, there is a listing of manufacturers in the appendices. Before the selection is made, you should obtain complete, up-to-date information from a number of these companies who will most likely supply you with a chart of which equipment they recommend for your specific situation and requirements. From this information, you will be able to effectively compare overall pricing and the abilities of several different Earth station packages. This will make the selection process much easier.

Once you have ordered the equipment, you can immediately begin to prepare the mounting site for the antenna. You should request the installation manual from your selected manufacturer in advance to allow you to utilize the time between equipment order and arrival to complete this phase of assembly. The sinking of the footers for the antenna base will require drilling three or more holes into the earth and pouring concrete. You can do this yourself or have a local contractor do the job for you. Either way, it will be necessary to allow about one week for the concrete to set. This time cannot be hurried and may even be lengthened, depending upon weather conditions in your area.

THE EQUIPMENT ARRIVES!

Finally, the glorious day is here, and your complete Earth station package arrives in several large boxes. Assuming that your site has been properly prepared in advance, your first job will be to examine each component carefully, looking for signs of damage. Since most antennas will arrive in sections that must be assembled, you will also want to check every bolt and bracket to make sure that you have not been accidentally shorted by the manufacturer. Most companies will include a packing list and a chart which will help you in taking this inventory. Damaged equipment should be reported immediately to the trucking company which made the delivery and simultaneously to the manufacturer. In most instances, it's best to wait until all equipment is in hand before beginning the construction. In a few instances, this may require waiting a few extra days; and while this is nerve-wracking, it will most likely assure a trouble-free installation which will serve you for many years to come.

You can probably assemble the antenna by yourself or with the aid of a friend. From a practical standpoint, however, three adults will make the assembly go much faster and more smoothly. These same three persons will definitely be needed to lift the antenna to its mounting position. You will also have to construct the base and bolt it to the concrete footers, and this is another job which is best done with some extra hands.

Before enlisting the aid of some fellow workers, sit down and re-read the entire assembly from cover to cover. Chances are, the manufacturer supplied you with this information as soon as you placed your order and you have already pored through it once. By re-reading the information, you will obtain a better understanding of just what is to be done and why certain specific procedures are necessary. A re-briefing the night before or the day of construction will put all of the assembly information fresh in your mind and you should be able to proceed without constant reference to the manual during each and every little step. After individual phases of construction are completed, a mark should be made in the manual indicating that this phase has been properly completed and tested. Most important is the alotting of an adequate amount of time to finish this project. Hurried installations often result in improper performance or a structure which may not be mechanically sound. This could mean that your antenna is blown into the next county by the first strong wind that comes along. Even though the manufacturer may state that the entire station may be built in a few hours, this may not be true in your case. It's a good idea to double the stated installation time to allow for any problems or inconsistencies which might crop up. For this reason I recommend that you begin your project early in the morning (at first light). If you start too late in the day, darkness may overtake you midway in the project and you will be forced to leave your partially assembled Earth station exposed to the elements overnight. If bad weather occurs the following day, your installation could be in real trouble. Anything partially assembled is more prone to accidental breakage and environmental damages than when completely installed. Also, if you have to stop midway, you can easily lose track of just where you left off when construction was halted. Most of the Earth stations discussed in this book are intended for one-day owner-installation.

Moving along in the assembly, the base is firmly anchored to the concrete footers in strict accordance with the manufacturer's instructions. This is most important, because most companies will not warranty their antennas against damage which is incurred through improper mounting. The manufacturers know best how to install their products for maximum stability and safety. This is an area where experimentation is not recommended on the part of the owner.

Now that the base is up, the antenna may be completely assembled. Often, this will involve the piecing together of the dish segments, which are sometimes referred to as petals. You will also have to properly align the pieces to make certain the dish follows the approval contours which will result in maximum efficiency. The manufacturer's instruction manual will completely explain how to perform this precise but relatively simple procedure.

After the dish portion is complete, the structure to which the feed horn is mounted will be attached. This will often involve three struts which are attached to the sides of the dish and which meet at the focal point of the antenna, forming a sort of tripod.

Before mounting the dish to the base, most installations call for the feed horn, LNA and polarity rotor to be mounted at the apex of the triangle mount. Associated coaxial cable and control wiring is also installed at this time. Some antenna designs will not use the tripod arrangement but instead will mount the LNA and feed horn at the end of a shaft which is connected at the center of the dish.

Once you have completed the antenna assembly, go back over all of your work, checking carefully with the specified installation procedures. If everything checks out, the dish is then aligned if this procedure is called for during this phase of construction. Now, it will be necessary for you and one or two other persons to lift the antenna and set it into place on its mount. Usually, two persons hold the antenna in place while the third inserts the connecting bolts at the back of the dish. These hold the entire assembly firmly to the mount.

Once the antenna is properly set up, it's a good idea to take a half hour break. During this period, you can go over the next stage of construction with your fellow workers. Make certain that you and your assistants understand exactly what is to be done. Good intentions coupled with a lack of knowledge have often been the ruin of many a project.

After everyone is refreshed, you can begin again. This next stage of construction is relatively simple and can probably be completed within an hour. It will vary slightly according to the particular components you are using, but it basically involves bolting the receiver to the antenna base, interconnecting the proper cables between this component and the LNA, and then running the feed lines and control wiring to the television set. If your receiver does not contain built-in modulator circuitry, a discrete unit will also be installed. Sometimes, this is done at the antenna site; but more often, the modulator will be mounted indoors near the television receiver and remote channel selector. You will want to be especially certain that all wiring connections are tight. This means tightening the collars of the coaxial connectors, tightening of all screw-in terminals, and the tight insertion of other types of control wiring connecting devices. If you live near the ocean, it would be a good idea to apply some waterproofing compound to connector surfaces in order to prevent corrosion. Be sure that there are no nicks in the protective outer cover of the coaxial cable. This can lead to water infiltration of the line. It's also a good idea to carefully examine the waterproof housings which are used to enclose receivers, LNAs, and other associated equipment which is to be mounted outdoors.

THE CHECKOUT

Once your TVRO Earth station is complete, go back through every major step of assembly in the manual and check to make certain all has gone according to the instructions. Sometimes, you will find that additional work is needed, such as tightening bolts, redoing connections, and readjusting coaxial cable. This procedure will also point out some areas which need a little more work, although not mentioned in the instruction manual. This could take the form of tying down coaxial cable and other conductors which could swing back and forth in a strong wind. It's usually possible, using waterproof masking tape, to tie these loose ends down in a few minutes. When cable is allowed to swing with the wind, mechanical breakdown can occur at the connectors. When the internal conductors are severed, reception will be lost.

After this close examination and any additional small jobs, you are ready to activate the system and check its performance. Antenna aiming information is included throughout this book and in the appendices. Direct your antenna toward a satellite. Turn your television receiver to the appropriate channel which will receive the output of the modulator (this is often channel 2 or 3 but can vary with modulator design). Using the satellite guide, select either vertical or horizontal polarization and switch the satellite receiver remote control unit to the appropriate channel. Now, activate the entire TVRO station, and you should receive a picture. It will probably be necessary to fine tune the audio control on the remote station for clearest reception. Ideally, the picture will be crisp, the audio clear, and the overall reception identical to what you can obtain from the standard television broadcast service.

Rarely, however, is ideal reception obtained in the first try. It will probably be necessary to do some minor realignment of the dish position at the antenna site using the manual controls. If you have motorized antenna controls, this can be done from your viewing site by utilizing the remote antenna control unit. Chances are, your first picture may be fuzzy or even nonexistent. Maneuvering the antenna just a bit should serve to clear up reception. Eventually, through trial and error adjustments, perfect reception will be obtained. As you become more and more familiar with your station and its various settings, tuning will be a simple matter which can be accomplished by any member of your family.

PROBLEMS

Let's assume that you try every antenna setting imaginable and still don't obtain adequate reception. Actually, if any portion of the TVRO Earth station is not operating properly, you will probably receive no picture at all. The first thing to do is to check the electrical connections to the station as a whole. In other words, make sure it's "plugged in". Most Earth station equipment contains lighted indicators which will tell you if individual components are receiving power. Start from the antenna and work back to the television receiver. Is the LNA receiving power? Chances are, this voltage source will be derived from the satellite receiver. Is the receiver activated? Is the modulator receiving power? A complete checkout procedure for each piece of equipment will most likely be provided by the manufacturer.

If you're sure that each component in the Earth station is receiving power, then additional checks are necessary. However, if one component is not being activated, it will be necessary to determine the cause of the problem. Are all cables which supply operating current properly connected? Have you observed connection polarity when separate lines are used for powering the LNA? A reversed connection here may possibly damage the internal circuitry of this device. An ac voltmeter will help you to determine if house current is available at the equipment which operates from this source. Check the fuses. Sometimes, receivers and other devices will blow a fuse for a variety of reasons. Replace any open fuses (making sure the power is off) and then try again. If the fuse blows again, then there may be internal circuit problems and the manufacturer should be consulted.

As was mentioned above, if your station is receiving power and you still aren't receiving a picture, you must follow another path of inspection. First of all, check all coaxial cable connections. These often are the cause of many TVRO problems especially if the cables are abused. Then too, coax often serves as the dc power line for the low noise amplifier. If all other equipment is properly activated and the LNA does not receive power, then this line should be highly suspect. In continuing your trace, inspect all other coaxial cables and any appropriate lines. If you're in doubt about the condition of any of these, check them out for continuity with an ohmmeter. Better yet, replace them entirely.

If a picture still is not received, check antenna alignment again and also make certain that the feed horn is positioned for the type of polarization used in the satellite transmission. If it's said to vertical polarization, switch it to horizontal just as a test. Remember, you will probably receive absolutely nothing if the wrong polarization is chosen.

If you still don't have reception, turn your antenna to another satellite and try again. It is conceivable that the first one is not operational for some reason, although this is also never the case. Also, try several different positions on the remote channel select unit. If your system is operational, you will certainly get something at the television receiver by trying a different satellite.

If all of this fails, then consult your buyer's manual once again to see if the manufacturer recommends any additional procedures. The checkout mentioned in this discussion is very broad in nature and specific Earth stations will come with manuals that go into a great deal more detail. If you can't determine what the problem is, then a call to the manufacturer will be necessary. Before doing this, however, go through your antenna alignment procedure one more time from the start. An error way back when you were choosing the site could mean that your dish is not aimed to the correct part of the sky. This can be a costly mistake, especially at this point in the setup, because your site location was probably chosen based upon these initial calculations. If you find you have made an error, point your antenna in the correct direction and start over again.

CALLING THE MANUFACTURER

Before calling the technical assistance department of the company from which you purchased your TVRO Earth station equipment, jot down all of the results of your troubleshooting checks. List which components seem to be receiving power or appear operational and those that aren't. Write down any specific questions you have about areas where you suspect the problem may be occurring. Check out your television receiver to make sure it is operational and note any unusual occurrences. Also, add to your list a description of any types of video or audio information you may be receiving on the satellite channels.

Before placing the call, have an assistant nearby. The technician may request that you try a certain procedure and report the results to him while on the phone. Your assistant may be able to help perform these additional checks or simply communicate what you find at the TVRO site, acting as a go-between. The companies I have been in contact with offer excellent technical assistance departments which are designed to take care of a multitude of problems by talking to the TVRO Earth station owner by telephone.

It should be pointed out that reputable companies check all of their equipment before it is sent out to the customer. Earth stations are not selling by the millions at the present time, so it is easier for the manufacturer to run comprehensive checks and to assure quality performance with a relatively small staff. Chances are, the equipment you receive (if not damaged in transit) will be fully operational and will require only minor adjustments after initial installation if any adjustments at all are actually necessary.

If from your report, the technician decides that a component is malfunctioning, it will probably be necessary for you to return it to them for service or replacement. Many companies will simply send out a replacement immediately in order to get your station operational in the shortest period of time.

In speaking with many TVRO Earth station owners, I have come across only one situation where a piece of defective equipment was the cause of a receiving failure upon initial installation and checkout. The manufacturer immediately sent out a replacement which arrived within 24 hours and was most apologetic about the inconvenience to their customer. Certainly, these situations can occur, but they are few and far between.

The vast majority of TVRO Earth station owners install their systems and receive some sort of reception on the first try. It is usually necessary to make some minor antenna adjustments and perhaps rearrange a cable or two to obtain perfect reception. The checkout procedure usually takes less than an hour for the first satellite received and just a few minutes for all those that follow.

INTERFERENCE

Some TVRO Earth station owners have experienced problems with interference to their reception by other communications companies which operate equipment that transmits a signal which can enter the TVRO system. Several manufacturers contacted mentioned that as a part of a normal sale, they will check to see what possible interference problems may exist for a particular location. The owner is then advised of the potential problems and supplied with information and/or filters which will serve to eliminate them. Of course, your television receiver may be subject to interference from CB radios, ignition noise from automobiles, and the other routine problems which have existed for a long time. A TVRO Earth station will often do nothing to prevent these problems from continuing and it will be necessary for the television owner to install the appropriate filters at the television receiver proper to clean this up. These can often be purchased at local electronic hobby stores.

SUMMARY

It is important to mention that most manufacturers of personal TVRO Earth stations designed for home assembly realize that the great majority of their customers are not scientists or technicians who have a great deal of expertise in microwave and space communications. The success of these manufacturers in marketing their products depends directly upon offering a package which can be installed in a short period of time by customers with average skills. When the installation is complete, the station should work perfectly. Should a package be offered which requires a great deal of technical skill to set up and align, it would not be marketable to the vast majority of average persons who are the prime targets of the TVRO Earth station industry.

By looking at installation plans from various manufacturers before ordering, the potential buyer can get a good idea of whether or not he feels qualified to perform the assembly and checkout procedures. These manuals are as important to the customer as is the station itself. Without a properly written manual, the joys of building your own TVRO Earth station can quickly turn into a massive headache. The step-by-step plans from the companies mentioned in this book can only be described as excellent.

When assembling your station, pay attention to every little detail. Follow the manufacturer's instructions religiously and don't take any shortcuts. If you do all of this, you can be 99% sure that you will quickly be the proud owner of a personal TVRO Earth station which will provide you with many years of information, entertainment, and untold satisfaction.

Appendix A
Satellite Video
Systems Complete Listing
of All Programming Sources
Available from the Satellites

RCA SATCOM 1

TR-1:	NICKELODEON—Premium Children's Programming
TR-2:	PTL (Praise the Lord)—Religious
TR-3:	WGN-TV, Chicago
TR-5:	THE MOVIE CHANNEL—24 hr./day new movies
TR-6:	WTBS, Atlanta—Ted Turner's Superstation
TR-7:	ESPN (Entertainment & Sports Network)—24 hr/day sports
TR-8:	CBN (Christian Broadcasting Network)—Religious
TR-9:	C-SPAN—Live coverage from the House of Representatives
	USA NETWORK: Madison Square Garden Sports, Calliope, and Black Entertainment Network
TR-10:	SHOWTIME (west)—first-run movies, entertainment specials
TR-11:	NICKELODEON—Premium Children's Programming
TR-12:	SHOWTIME (east)—first-run movies, entertainment specials
	ABC Network Special Programming
TR-13:	TBN (Trinity Broadcasting Network)—Religious
TR-14:	CNN (Cable News Network)—24 hr/day news
TR-16:	SHOWTIME (spare)
	Occasional network, remote and sports events feeds
	ACSN (Appalachian Community Service Network)
	WWS (Window on Wall Street)—financial
TR-17:	WOR-TV, New York
TR-18:	REUTERS NEWS SERVICE
	GALAVISION—The best in Spanish-oriented programming

TR-20:	HOME BOX OFFICE CINEMAX (east)—time-structured HBO
TR-21:	HTN (Home Theater Network)—quality G & PG movies
TR-22:	HBO (Home Box Office) (west)—first-run movies, entertainment specials
	MSN (Modern Satellite Network)—general entertainment
TR-23:	HBO CINEMAX (west)—time-structured HBO
TR-24:	HBO (east)—first-run movies, entertainment specials

Audio Services on SATCOM 1

TR-2:	WAME (AM), Charlotte, North Carolina (6.2)
TR-3:	WFMT (FM), Chicago, Illinois (5.8)
	Seeburg Easy Listening Music (7.6)

ATT/GTE COMSTAR 2

TR-2:	Occassional transmissions: teleconferencing, sporting events, news & network feeds
TR-4:	Occasional transmissions: teleconferencing, sporting events, news & network feeds
TR-6:	Occasional transmissions: teleconferencing, sporting events, news & network feeds
TR-7:	NCN (National Christian Network)—religious
	BRAVO—Performing and Cultural Arts Programming
	ESCAPADE—"R" rated movies only
TR-10:	Occasional transmissions: teleconferencing, sporting events, news & network feeds
TR-13:	TBN (Trinity Broadcasting Network)—religious
TR-15:	Occasional transmissions: teleconferencing, news, and network feeds
TR-17:	HOME BOX OFFICE CINEMAX (east)—time-structured HBO
TR-18:	HOME BOX OFFICE (east)—first-run movies, entertainment specials
TR-19:	LVEN (Las Vegas Entertainment Network)—Live on-stage programming from Las Vegas
	CINEAMERICA—entertainment programming directed for ages over 50
TR-21:	Occasional transmissions: teleconferencing, sporting events, news & network feeds
TR-22:	Occasional transmissions: teleconferencing, sporting events, news & network feeds

Audio Services on COMSTAR 2

| TR-7: | Family Radio Network (east) (5.8) |
| | Family Radio Network (west) (7.7) |

WU WESTAR 3

TR-2:   HUGES SPORTS NETWORK
Occasional transmissions: sporting events, news & network feeds

TR-3:   XEW-TV, Mexico City

TR-5:   WOLD COMMUNICATIONS—occasional transmissions: sporting events, news & network feeds
PRIVATE SCREENINGS—Hardcore, sexploitation "R"

TR-6:   CBS Network Contract Channel—live network feeds
Occasional transmissions: sporting events, news news feeds

TRS-8:   SIN (Spanish International Network)

TR-9:   SPN (Satellite Program Network)—classic movies

TR-10:   ABC Network Contract Channel—live network feeds

TR-11:   CNN (Cable News Network) Contract Channel—news feeds
Occasional transmissions: sporting events

TR-12:   Occasional transmissions: sporting events, news & network feeds

WU WESTAR 1

TR-1:   Occasional transmissions: sporting events, news & network feeds

TR-3:   Occasional transmissions: sporting events, news & network feeds

TR-5:   Occasional transmissions: sporting events, news & network feeds
VEU (Video Entertainment Unlimited)—STV feed: first-run movies, concert specials, sporting events

TR-6:   Occasional transmissions: sporting events, news & network feeds

TR-8:   PBS (Public Broadcasting) Schedule A Programming

TR-9:   PBS (Public Broadcasting) Schedule B Programming

TR-11:   PBS (Public Broadcasting) Schedule C Programming

TR-12:   PBS (Public Broadcasting) Occasional Feeds
CENTRAL EDUCATIONAL NETWORK
Occasional transmissions: sporting events, news & network feeds

WU WESTAR 2

TR-2:   Occasional transmissions: sporting events, news & network feeds

TR-9:   Occasional transmissions: sporting events, news & network feeds

RCA SATCOM 2

TR-2:	Occasional transmissions: sporting events, news & network feeds
TR-5:	Occasional transmissions: sporting events, news & network feeds
TR-8:	NBC Network Contract Channel—live & taped network feeds
TR-9:	ALASKAN FORCES SATELLITE NETWORK— assorted independent & network programming
TR-13:	NASA CONTRACT CHANNEL
TR-23:	ALASKA SATELLITE TELEVISION PROJECT— assorted network and independent programming

ATT/GTE COMSTAR 1

TR-20:	Occasional transmissions: sporting events, news & network feeds

ANIK B(2) (CAMADOAM)

TR-4:	Occasional transmissions: sporting events, news & CBC/CTV network feeds
TR-6:	CBC NORTH—assorted CBC network programming
TR-7:	Occasional Transmissions: sporting events, news — CBC/CYV network feeds
TR-8:	CBC (French Channel)—French language CBC programming
TR-9:	CBC (English Channel-1) English CBC programming
TR-10:	CBC (English Channel-2) English CBC programming

ANIK 3 (CANADIAN)

TR-1:	Occasional Transmissions: sporting events, news & CBC/CTV network feeds
TR-4:	Daily Live Coverage of Canadian House of Commons from Ottawa (with French translation)
TR-7:	Daily Live Coverage of the Canadian House of Commons from Ottawa (standard English)
TR-12:	CTV NORTH—assorted CTV network programming

(Courtesy Gillaspie and Associates, Inc.)

Appendix B
List of Geostationary Space
Stations by Orbital Positions

The following list includes both satellites already in orbit and those planned for future launching into the geostationary satellite orbit.

The table is based on, and limited to, information supplied to the International Frequency Registration Board (IFRB) by ITU Member administrations under the provisions of the Radio Regulations paragraphs 639AA, 639AJ, 639BA. The designations of the satellites are those officially notified and may not always correspond to the name in general use.

Orbital Position			Space station	GHz	<1	<3	4	6	7	11	12	14	>15
0	E		F/GEO	GEOS-2		1	3						
0	E		F/MET	METEOSAT		1	3						
10	E		F/OTS	OTS		1					11		14
13	E	#	F	EUTELSAT 1-2		1					11		14
17	E	#	ARS	SABS							11	12	14
19	E	*	ARS	ARABSAT I			3	4	6				
26	E	*	ARS	ARABSAT II			3	4	6				
26	E	*	IRN	ZOHREH-2							11		14
29	E		F/GEO	GEOS-2		1	3						
34	E	*	IRN	ZOHREH-1							11	12	14
35	E		URS	STATSIONAR-2				4	6				
40	E	*	F/MRS	MARECS-D		1	3	4	6				
40	E		F/MTS	MAROTS		1	3						
40	E	#	URS	STATSIONAR-12				4	6				
45	E	*	URS	GALS-2						7			
45	E	*	URS	LOUTCH P2							11		14
45	E		URS	STATSIONAR-9				4	6				
45	E	*	URS	VOLNA-3		1	3						
47	E	*	IRN	ZOHREH-3							11		14
53	E		URS	LOUTCH-2							11		14
53	E		URS/IK	STATSIONAR-5				4	6				
53	E		URS	VOLNA-4			3						
56.5	E		USA/IT	INTELSAT3 INDN1				4	6				
57	E	*	USA/IT	INTELSAT4 INDN2				4	6				
57	E	*	USA/IT	INTELSAT4A INDN2				4	6				
57	E	#	USA	INTELSAT5 INDN3				4	6		11		14
57	E	#	USA	INTELSAT MCS INDN C 3				4	6				
60	E		USA/IT	INTELSAT4 INDN2				4	6				
60	E		USA/IT	INTELSAT4A INDN2				4	6				
60	E	*	USA/IT	INTELSAT5 INDN2				4	6		11		14
60	E	*	USA/IT	INTELSAT MCS INDN B 3				4	6				
60	E		USA	USGCSS PHASE2 INDN						7			
60	E	#	USA	USGCSS PHASE3 INDN						7			
63	E		USA/IT	INTELSAT4 INDN1				4	6				
63	E		USA/IT	INTELSAT4A INDN1				4	6				
63	E	*	USA/IT	INTELSAT5 INDN1				4	6		11		14
63	E	*	USA/IT	INTELSAT MCS INDN A 3				4	6				
64.5	E	*	F/MRS	MARECS-C		1	3	4	6				
66	E	*	USA/IT	INTELSAT4 INDN1				4	6				
66	E	*	USA/IT	INTELSAT4A INDN1				4	6				
66	E	#	USA	INTELSAT5 INDN4				4	6		11		14
66	E	#	USA	INTELSAT MCS INDN D 3				4	6				
73	E		USA	MARISAT-INDN		1	3*	4*	6*				
74	E		IND	INSAT-1A		1	3	4	6				

Orbital Position			Space station	GHz <1	<3	4	6	7	11	12	14	>15
75 E		USA	FLTSATC INDN	1				7				
77 E		INS	PALAPA-2			4	6					
80 E		URS	STATSIONAR-1			4	6					
80 E	#	URS	STATSIONAR-13			4	6					
83 E		INS	PALAPA-1			4	6					
85 E	*	URS	GALS-3					7				
85 E	*	URS	LOUTCH P3						11		14	
85 E		URS	STATSIONAR-3			4	6					
85 E	*	URS	VOLNA-5	1	3							
90 E		URS	LOUTCH-3						11		14	
90 E		URS	STATSIONAR-6			4	6					
90 E	#	URS	VOLNA-8		3							
94 E	*	IND	INSAT-1B	1	3	4	6					
95 E	#	URS	STATSIONAR-14			4	6					
99 E		URS	STATSIONAR-T	1			6					
99 E	#	URS	STATSIONAR-T2	1			6					
102 E	*	IND	ISCOM	1		4	6					
108 E	*	INS	PALAPA-B1			4	6					
110 E		J	BSE		3						14	
113 E	*	INS	PALAPA-B2			4	6					
118 E	*	INS	PALAPA-B3			4	6					
125 E	*	CHN	STW-1			4	6					
130 E		J	ETS-2	1	3				11			34
130 E	#	J	CS-2A		3	4	6					20/30
130 E	#	URS	STATSIONAR-15			4	6					
135 E	#	J	CS-2B		3	4	6					20/30
135 E		J	CSE		3	4	6					18/29
140 E	*	J	GMS	1	3							
140 E		URS	LOUTCH-4						11		14	
140 E		URS	STATSIONAR-7			4	6					
140 E		URS	VOLNA-6		3							
172 E		USA	FLTSATC W PAC	1				7				
174 E		USA/IT	INTELSAT4 PAC1			4	6					
174 E	*	USA/IT	INTELSAT4A PAC1			4	6					
174 E	#	USA	INTELSAT5 PAC1			4	6		11		14	
175 E		USA	USGCSS PHASE2 W PAC					7				
175 E	#	USA	USGCSS PHASE3 W PAC					7				
176.5 E		USA	MARISAT-PAC	1	3	4	6					
179 E		USA/IT	INTELSAT4 PAC2			4	6					
179 E	*	USA/IT	INTELSAT4A PAC2			4	6					
179 E	#	USA	INTELSAT5 PAC2			4	6		11		14	
172 W	*	F/MRS	MARECS-B	1	3	4	6					
170 W	*	URS	GALS-4					7				
170 W	*	URS	LOUTCH P4						11		14	
170 W		URS	STATSIONAR-10			4	6					
170 W	*	URS	VOLNA-7	1	3							
149 W		USA	ATS-1	1		4	6					
136 W		USA	US SATCOM-1			4	6					
135 W		USA	GOES WEST	1	3							
135 W	#	USA	USGCSS PHASE3 E PAC					7				
135 W		USA	USGCSS PHASE2 E PAC					7				
132 W	#	USA	US SATCOM-3			4	6					
128 W		USA	COMSTAR D1			4	6					
123.5 W		USA	WESTAR-2			4	6					
122 W		USA	USASAT-6A						11		14	
119 W		USA	US SATCOM-2			4	6					
116 W	*	CAN	ANIK-C2						11		14	
114 W		CAN	ANIK-A3			4	6					
112.5 W	*	CAN	ANIK-C1						11		14	
109 W		CAN	ANIK-A2			4	6					
109 W		CAN	ANIK-B1			4	6		11		14	
106 W		USA	USASAT-6B						11		14	
105 W		USA	ATS-5	1	3							
104 W	#	CAN	TELESAT D-1			4	6					

Orbital Position		Space station		Frequency Bands GHz	<1	<3	4	6	7	11	12	14
104 W		CAN	ANIK-A1				4	6				
102 W	#	MEX	SATMEX-1				4	6				
100 W		USA	FLTSATC E PAC		1				7			
99 W		USA	WESTAR-1				4	6				
95 W		USA	COMSTAR D2				4	6				
91 W	*	USA	WESTAR-3				4	6				
87 W		USA	COMSTAR D3				4	6				
86 W		USA	ATS-3		1							
75.4 W		CLM	SATCOL-2				4	6				
75 W		USA	GOES EAST		1	3						
75 W		CLM	SATCOL-1				4	6				
34.5 W	#	USA	INTELSAT MCS ATL E			3	4	6				
34.5 W		USA/IT	INTELSAT4 ATL5				4	6				
34.5 W		USA/IT	INTELSAT4A ATL4				4	6				
34.5 W	*	USA/IT	INTELSAT5 ATL4				4	6		11		14
31 W	*	USA/IT	INTELSAT4A ATL4				4	6				
29.5 W		USA/IT	INTELSAT4 ATL2				4	6				
29.5 W		USA/IT	INTELSAT4A ATL3				4	6				
29.5 W		USA/IT	INTELSAT5 ATL3				4	6		11		14
27.5 W	*	USA/IT	INTELSAT4A ATL3				4	6				
27.5 W	*	USA/IT	INTELSAT5 ATL3				4	6		11		14
27.5 W	*	USA/IT	INTELSAT MCS ATL B			3	4	6				
25 W	*	URS	GALS-1						7			
25 W	*	URS	LOUTCH P1							11		14
25 W		URS	STATSIONAR-8				4	6				
25 W	*	URS	VOLNA-1		1	3						
24.5 W	#	USA	INTELSAT MCS ATL D			3	4	6				
24.5 W		USA/IT	INTELSAT4A ATL1				4	6				
24.5 W		USA/IT	INTELSAT5 ATL1				4	6		11		14
23 W		USA	FLTSATC ATL		1				7			
21.5 W	#	USA	INTELSAT MCS ATL C			3	4	6				
21.5 W	#	USA	INTELSAT5 ATL5				4	6		11		14
21.5 W	*	USA/IT	INTELSAT4 ATL2				4	6				
21.5 W	*	USA/IT	INTELSAT4A ATL1				4	6				
19.5 W		USA/IT	INTELSAT4 ATL3				4	6				
19.5 W		USA/IT	INTELSAT4A ATL2				4	6				
18.5 W	*	USA/IT	INTELSAT4 ATL3				4	6				
18.5 W	*	USA/IT	INTELSAT4A ATL2				4	6				
18.5 W	*	USA/IT	INTELSAT5 ATL2				4	6		11		14
18.5 W	*	USA/IT	INTELSAT MCS ATL A			3	4	6				
18 W	*	BEL	SATCOM III ATL						7			
18 W	*	BEL	SATCOM-III						7			
18 W		BEL	SATCOM-II						7			
15 W	*	F/MRS	MARECS-A		1	3	4	6				
15 W		I	SIRIO		1					11		
15 W		USA	MARISAT-ATL		1	3	4	6				
14 W		URS	LOUTCH-1							11		14
14 W		URS/IK	STATSIONAR-4				4	6				
14 W		URS	VOLNA-2			3						
13 W		USA	USGCSS PHASE2 ATL						7			
12 W	*	USA	USGCSS PHASE2 ATL						7			
12 W	#	USA	USGCSS PHASE2 ATL						7			
11.5 W		F/SYM	SYMPHONIE-2		1		4	6				
11.5 W		F/SYM	SYMPHONIE-3		1		4	6				
10 W	#	F	TELECOM-1A			3	4	6	7		12	14
8.5 W	#	URS	STATSIONAR-11				4	6				
7 W	#	F	TELECOM-1B			3	4	6	7		12	14
4 W		USA/IT	INTELSAT4 ATL1				4	6				
1 W		USA/IT	INTELSAT4 ATL4				4	6				

* Under co-ordination RR639AJ

Advanced publication only under RR639AA

(courtesy of International Telecommunication Union)

Appendix C
Table of Artificial Satellites Launched in 1980

Code name / Spacecraft description	International number	Country / Organization / Site of launching	Date	Perigee / Apogee	Period / Inclination	Frequencies / Transmitter power	Observations
Cosmos-1149	1980-1-A	USSR (PLE)	9 Jan.	208 km / 414 km	90.4 min / 70.29°		Photographic reconnaissance satellite. Recovered on 23 January 1980
46th Molnya-1 hermetically sealed cylinder with conical ends. width: 1000 kg, 6 solar panels	1980-2-A	USSR (PLE)	11 Jan.	478 km / 40 830 km	737 min / 62.8°	800 MHz band 40 W (emission) 1000 MHz band (reception) 3400-4100 MHz (retransmission of television)	Apparatus for transmitting television programmes and multichannel radiocommunications
Cosmos-1150	1980-3-A	USSR (PLE)	14 Jan.	989 km / 1028 km	105.0 min / 83.0°		Navigation satellite
FLTSATCOM-3 3-axis stabilized hexagonal satellite. width: 2.44 m, overall height: 6.70 m, mass at launch: 1875 kg, mass in orbit: 1005 kg	1980-4-A	United States USDOD (ETR)	18 Jan.	35 745 km / 35 829 km; in geostationary orbit at 23° W	1436.1 min / 2.6°	240-400 MHz band (communications) 2252.5 - 2262.5 MHz 2.4 W (telemetry)	Government communications satellite. Replaces FLTSATCOM-2 which is being moved to 75° E. Nine 25 kHz channels and twelve 5 kHz channels for small mobile users. A 25 kHz broadcast channel and a 500 kHz channel
Cosmos-1151	1980-5-A	USSR (PLE)	23 Jan.	650 km / 678 km	97.8 min / 82.5°		Ocean monitoring satellite
Cosmos-1152	1980-6-A	USSR (PLE)	24 Jan.	181 km / 370 km	89.7 min / 67.1°		High-resolution, reconnaissance satellite. Recovered on 6 February 1980
Cosmos-1153	1980-7-A	USSR (PLE)	25 Jan.	983 km / 1031 km	105 min / 83°		Navigation satellite
Cosmos-1154	1980-8-A	USSR (PLE)	30 Jan.	634 km / 671 km	97.3 min / 81.3°		Electronic monitoring satellite
Cosmos-1155	1980-9-A	USSR (PLE)	7 Feb.	206 km / 422 km	90.4 min / 72.9°		Medium-resolution reconnaissance satellite. Recovered on 21 February 1980
KH-11	1980-10-A	United States USAF (WTR)	7 Feb.	220 km / 498 km	91.7 min / 97.0°		Digital imaging reconnaissance satellite

ITU Telecommunication Journal

Code name Spacecraft description	International number	Country Organization Site of launching	Date	Perigee Apogee	Period Inclination	Frequencies Transmitter power	Observations
Navstar-5	1980-11-A	United States (WTR)	9 Feb.	20 095 km 20 165 km	715.9 min 63.7°		Global positioning system navigation satellite. Replaces Navstar-1
Cosmos-1156 to **Cosmos-1163** mass: 40 kg each	1980-12-A to 1980-12-H	USSR (PLE)	12 Feb.	1450 km 1528 km	115.4 min 74.0°		Government communication satellites
Cosmos-1164	1980-13-A	USSR (PLE)	12 Feb.	220 km 640 km	92.9 min 62.8°		
SMM 3-axis stabilized satellite: width: 1.20 m; length: 4 m; mass: 2315 kg; 2 fixed solar arrays (3 kW): Ni-Cd batteries	1980-14-A	United States NASA (ETR)	14 Feb.	571.5 km 573.5 km	96.12 min 28.5°	2287.5 MHz (tracking and telemetry)	Solar Maximum Mission. Objectives: to measure solar radiation during the period of maximum solar activity. Carries gamma ray spectrometer, hard X-ray burst spectrometer, hard X-ray imaging spectrometer, ultraviolet spectrometer and polarimeter, X-ray polychromter, coronograph/polarimeter and solar constant monitoring package
Tansei-4 (MS-T4)	1980-15-A	Japan ISAS (KSC)	17 Feb.	517 km 672 km	96.5 min 38.7°	136.725 MHz (tracking) 400.45 ; 2280.5 MHz (telemetry)	Tests of new technology for future satellites and test of the new M-3S launcher
Raduga-5 3-axis stabilized satellite; mass: 5 tonnes; solar cells	1980-16-A	USSR (BAI)	20 Feb.	36 610 km geosynchronous orbit	24 h 38 min 0.4°	5.7-6.2 GHz (reception) 3.4-3.9 GHz (emission)	Carries apparatus for transmitting television programmes and multichannel radiocommunications
Cosmos-1165	1980-17-A	USSR (PLE)	21 Feb.	182 km 379 km	89.8 min 72.9°		High-resolution reconnaissance satellite. Recovered on 5 March 1980

Code name Spacecraft description	International number	Country Organization Site of launching	Date	Perigee Apogee	Period Inclination	Frequencies Transmitter power	Observations
Ayame-2 cylindrical satellite: diameter: 1 m; height: 1.50 m; mass: 260 kg	1980-18-A	Japan NSDA (TSC)	22 Feb.	206.9 km 35 512 km	625.8 min 24.6°	31.65 GHz 3.2 W 4.075; 4.080 GHz 4.7 W 3.940 GHz 3.5 W 136.112 MHz 2 or 8 W	Experimental telecommunication satellite. Was intended for geostationary orbit but contact was lost while still in transfer orbit
No name	1980-19-A to 1980-19-C	United States USN (WTR)	3 March	1053 km 1151 km	107.1 min 63.5°		Ocean surveillance satellite system. Three satellites
Cosmos-1166	1980-20-A	USSR (PLE)	4 March	208 km 406 km	90.3 min 72.9°		Medium-resolution photographic reconnaissance satellite. Recovered on 18 March 1980
Cosmos-1167	1980-21-A	USSR (PLE)	14 March	438 km 457 km	93.3 min 65°		Ocean surveillance satellite
Cosmos-1168	1980-22-A	USSR (PLE)	17 March	981 km 1026 km	104.9 min 82.9°		Navigation satellite
Cosmos-1169	1980-23-A	USSR (PLE)	27 March	478 km 521 km	94.5 min 65.8°		Satellite intercept programme
Progress-8 modified *Soyuz* spacecraft without the descent section; mass at launch: 7 tonnes	1980-24-A	USSR (BAI)	27 March	192 km 266 km	88.8 min 51.6°		Expendable supply craft. Docked with *Salyut-6* on 29 March. Separated on 25 April and was deorbited over the Pacific Ocean on 26 April 1980
Cosmos-1170	1980-25-A	USSR (BAI)	1 April	181 km 386 km	89.9 min 70.4°		High-resolution photographic reconnaissance satellite. Recovered on 13 April 1980
Cosmos-1171	1980-26-A	USSR (PLE)	3 April	976 km 1017 km	105 min 65.8°		Satellite intercept programme. Target vehicle for *Cosmos-1174*
Soyuz-35 3-part spacecraft: 2 spherical habitable modules (orbital compartment and command module) connected in tandem to a cylindrical service module: diameter: 2.70 m; height: 7.10 m; mass: 6.7 tonnes; 2 solar arrays	1980-27-A	USSR (BAI)	9 April	276 km 315 km	90.3 min 51.6°		Two-man spacecraft: L. Popov, commander; V. Ryumin, flight engineer. Docked with *Salyut-6* on 10 April. Returned to Earth carrying *Soyuz-36* cosmonauts on 3 June 1980, landing some 440 km north-east of Baikonur Cosmodrome

Satellite	Designation	Country (site)	Date	Altitude	Period / Inclination	Description
Cosmos-1172	1980-28-A	USSR (PLE)	12 April	637 km / 40 160 km	726 min / 62.8°	Early warning satellite
Cosmos-1173	1980-29-A	USSR (BAI)	17 April	180 km / 379 km	89.9 min / 70.3°	High-resolution photographic reconnaissance satellite. Similar to *Cosmos-1170*. Recovered on 28 April 1980
Cosmos-1174	1980-30-A	USSR (BAI)	18 April	387 km / 1035 km	98.6 min / 65.8°	Satellite intercept programme. Interceptor vehicle for *Cosmos-1171* target vehicle. The test was a failure. *Cosmos-1174* was exploded in space on 20 April 1980
Cosmos-1175	1980-31-A	USSR (PLE)	18 April	317 km / 485 km	92.3 min / 62.5°	Satellite intercept programme Interceptor vehicle employing an optical-thermal guidance system. Decayed on 28 May 1980
Navstar-6	1980-32-A	United States (WTR)	26 April	19 622 km / 20 231 km	707.6 min / 62.9°	Global positioning system navigation satellite
Progress-9 modified *Soyuz* spacecraft without the descent section; mass at launch: 7 tonnes	1980-33-A	USSR (BAI)	27 April	192 km / 275 km	88.9 min / 51.6°	Expendable supply craft. Docked with *Salyut-6* on 29 April, undocked on 20 May and was made to re-enter the Earth's atmosphere on 22 May 1980
Cosmos-1176	1980-34-A	USSR	29 April	260 km / 265 km	89.6 min / 65.0°	Ocean surveillance satellite. Carries a nuclear reactor. Similar to *Cosmos-954*
Cosmos-1177	1980-35-A	USSR (PLE)	29 April	181 km / 365 km	89.7 min / 67.2°	High-resolution photographic reconnaissance satellite. Recovered on 12 June 1980
Cosmos-1178	1980-36-A	USSR (PLE)	7 May	207 km / 417 km	90.4 min / 72.9°	Reconnaissance satellite. Recovered on 22 May 1980
Cosmos-1179	1980-37-A	USSR (PLE)	14 May	310 km / 1570 km	103.5 min / 83.0°	Navigation satellite
Cosmos-1180	1980-38-A	USSR (PLE)	15 May	240 km / 296 km	89.8 min / 62.8°	Satellite for geophysical observations and measurements. Recovered on 25 May 1980

Code name Spacecraft description	International number	Country Organization Site of launching	Date	Perigee Apogee	Period Inclination	Frequencies Transmitter power	Observations
Cosmos-1181	1980-39-A	USSR	20 May	992 km 1020 km	105 min 82°		Navigation satellite
Cosmos-1182	1980-40-A	USSR (PLE)	23 May	221 km 278 km	89.2 min 82.3°		Medium-resolution photographic reconnaissance satellite. Recovered on 5 June 1980
Soyuz-36 3-part spacecraft: 2 spherical habitable modules (orbital compartment and command module) connected in tandem to a cylindrical service module; diameter: 2.70 m: height: 7.10 m; mass: 6.7 tonnes; 2 solar arrays	1980-41-A	USSR (BAI)	26 May	198 km 216 km	88.0 min 51.6°		Two-man spacecraft. V. Kubasov, commander: B. Farkas (Hungary), research cosmonaut. Docked with Salyut-6 on 28 May. Soyuz-36 was returned to Earth with Soyuz-37 cosmonauts aboard on 31 July 1980
Cosmos-1183	1980-42-A	USSR	28 May	208 km 414 km	90.4 min 72.9°		Medium-resolution photographic reconnaissance satellite. Recovered on 11 June 1980
NOAA-B	1980-43-A	United States NOAA (WTR)	29 May	273 km 1453 km	102.2 min 92.3°		Owing to malfunction of the launch vehicle, proper orbit was not attained and the spacecraft is considered inoperable
Cosmos-1184	1980-44-A	USSR (PLE)	4 June	621 km 662 km	97.4 min 81.2°		Electronic monitoring satellite
Soyuz-T2 solar batteries	1980-45-A	USSR (BAI)	5 June	267 km 316 km	90.25 min 51.6°		Two-man spacecraft. Y. Malishev and V. Aksenov, cosmonauts. Docked with accessories port of Salyut-6 on 6 June. Recovered on 9 June 200 km south-east of Dzhezkazgan
Cosmos-1185	1980-46-A	USSR (PLE)	6 June	226 km 308 km	89.5 min 82.3°		Medium-resolution photographic reconnaissance satellite. Recovered on 20 June 1980
Cosmos-1186	1980-47-A	USSR (PLE)	6 June	473 km 519 km	94.5 min 74.0°		Electronic monitoring satellite

Name	Designation	Country	Date	Altitude	Period / Inclination	Frequency	Remarks
		(PLE)	… June	210 km 332 km	72.9 min 89.6°		Medium-resolution photographic reconnaissance satellite. Recovered on 26 June 1980
Gorizont-4 3-axis stabilized spacecraft	1980-49-A	USSR (BAI)	14 June	36 515 km	24 h 33 min 0.8°	3.4-3.9 GHz (emission) 5.7-6.2 GHz (reception)	Communication satellite for transmission of telegraph and telephone messages and for transmission of television programmes
Cosmos-1188	1980-50-A	USSR (PLE)	14 June	628 km 40 165 km	726 min 62.8°		Early warning satellite
30th Meteor-1 3-axis stabilized cylindrical satellite mass: 2200 kg.; sun-oriented solar panels	1980-51-A	USSR CAHS (PLE)	18 June	589 km 678 km	97.3 min 98.0°		Meteorological satellite
Big Bird	1980-52-A	United States USAF (WTR)	18 June	165 km 254 km	88.5 min 96.5°		Reconnaissance satellite
No name	1980-52-C	United States	18 June	1325 km 1329 km	112.2 min 96.6°		
47th Molnya-1 hermetically-sealed cylinder with conical ends: mass: 1000 kg.: 6 solar panels	1980-53-A	USSR (PLE)	21 June	658 km 40 707 km	738 min 62.5°	800 MHz band 40 W (emission) 1000 MHz band (reception) 3400-4100 MHz (retransmission of television)	Television and multichannel radiocommunications
Cosmos-1189	1980-54-A	USSR (PLE)	26 June	209 km 330 km	89.5 min 72.9°		Medium-resolution photographic reconnaissance satellite. Recovered on 10 July 1980
Progress-10 modified *Soyuz* spacecraft without the descent section: mass at launch: 7 tonnes	1980-55-A	USSR	29 June	191 km 281 km	88.9 min 51.6°		Expendable supply craft. Docked with *Salyut-6* on 1 July and was made to re-enter the Earth's atmosphere on 19 July 1980

Code name Spacecraft description	International number	Country Organization Site of launching	Date	Perigee Apogee	Period Inclination	Frequencies Transmitter power	Observations
Cosmos-1190	1980-56-A	USSR (PLE)	1 July	792 km 829 km	100.8 min 74.0°		Electronic monitoring satellite
Cosmos-1191	1980-57-A	USSR (PLE)	2 July	646 km 40 165 km	726 min 62.8°		Early warning satellite
Cosmos-1192 to Cosmos-1199 mass: 40 kg each	1980-58-A to 1980-58-H	USSR (PLE)	9 July	1451 km 1522 km	115.3 min 74.0°		Government communication satellites
Cosmos-1200	1980-59-A	USSR (PLE)	9 July	209 km 332 km	89.5 min 72.9°		Medium-resolution photographic reconnaissance satellite. Recovered on 23 July 1980
Ekran-5 (Statsionar) 3-axis stabilized satellite: mass: 5 tonnes: solar cells	1980-60-A	USSR (BAI)	14 July	34 474 km geostationary orbit	1420 min 0.36°	5.7-6.2 GHz (reception) 3.4-3.9 GHz (emission)	Television relay satellite
Cosmos-1201	1980-61-A	USSR (PLE)	15 July	220 km 274 km	89.1 min 82.3°		Natural resources satellite. Recovered on 28 July 1980
Rohini-1 mass: 35 kg	1980-62-A	India (SSC)	18 July				First satellite launched by Indian SLV-3 solid propellant 4-stage rocket system
13th Molnya-3 3-axis stabilized satellite: mass: 1500 kg	1980-63-A	USSR (PLE)	18 July	467 km 40 815 km	736 min 62.8°	5.9-6.2 GHz (reception) 3.6-3.9 GHz (emission)	Television and multichannel radiocommunications
Soyuz-37 3-part spacecraft: 2 spherical habitable modules (orbital compartment and command module) connected in tandem to a cylindrical service module: diameter: 2.70 m: height: 7.10 m: mass: 6.7 tonnes: 2 solar arrays	1980-64-A	USSR (BAI)	23 July	263 km 312 km	90.0 min 51.6°		Two-man spacecraft: cosmonaut V. Gorbatko and cosmonaut-researcher Fam Tuan (Viet Nam). Docked with Salyut-6 on 24 July. Soyuz-37 cosmonauts returned to Earth aboard Soyuz-36 on 31 July 1980. Soyuz-37 spacecraft was returned to Earth with Soyuz-35 cosmonauts Popov and Ryumin on 11 October 1980

Name	Int'l designation	Country (agency)	Date	Orbit	Period / inclination	Frequency / power	Remarks
Cosmos-1202	1980-65-A	USSR (PLE)	25 July	209 km 333 km	89.6 min 72.9°		Medium-resolution photographic reconnaissance satellite. Recovered on 7 August 1980
Cosmos-1203	1980-66-A	USSR (PLE)	31 July	227 km 303 km	89.5 min 82.3°		Recovered on 14 August 1980
Cosmos-1204	1980-67-A	USSR (AKY)	31 July	346 km 546 km	93.3 min 50.7°		
Cosmos-1205	1980-68-A	USSR (PLE)	12 Aug.	208 km 332 km	89.6 min 72.8°		Photographic reconnaissance satellite. Recovered on 26 August 1980
Cosmos-1206	1980-69-A	USSR (PLE)	15 Aug.	630 km 659 km	97.4 min 81.2°		Electronic monitoring satellite
Cosmos-1207	1980-70-A	USSR (PLE)	22 Aug.	218 km 282 km	89.2 min 82.3°		Film-return earth resources satellite. Recovered on 4 September 1980
Cosmos-1208	1980-71-A	USSR	26 Aug.	181 km 362 km	89.6 min 67.1°		Long-duration reconnaissance satellite. Recovered on 24 September 1980
Cosmos-1209	1980-72-A	USSR (PLE)	3 Sept.	222 km 306 km	89.4 min 82.3°		Earth resources satellite. Recovered on 17 September 1980
6th Meteor-2	1980-73-A	USSR CAHS (PLE)	9 Sept.	868 km 906 km	102.4 min 81.2°	137.3 MHz 5 W (APT)	Meteorological satellite. Scanning telephotometer and television-type scanning equipment (0.5 to 0.7 μm), infrared scanning radiometer (8 to 12 μm)
GOES-4 cylindrical spin-stabilized satellite: diameter: 1.90 m; height: 2.30 m; mass: 397 kg	1980-74-A	United States (ETR)	9 Sept.	34 264 km 49 830 km	1767 min 0.25° in geosynchronous orbit at 95° W	2209 MHz: 2214 MHz (telemetry)	Geostationary Operational Environmental Satellite. Carries a visible and infrared spin-scan radiometer (VISSR) to provide data on the vertical structures of temperature and moisture in the atmosphere

Code name / Spacecraft description	International number	Country Organization Site of launching	Date	Perigee Apogee	Period Inclination	Frequencies Transmitter power	Observations
Soyuz-38 3-part spacecraft: 2 spherical habitable modules (orbital compartment and command module) connected in tandem to a cylindrical service module; diameter: 2.70 m; height: 7.10 m; mass: 6680 kg; 2 solar arrays	1980-75-A	USSR (BAI)	18 Sept.	199 km 273 km	88.9 min 51.6°		Two-man spacecraft: Y. V. Romanenko, flight commander; A. Tomayo Méndez (Cuba). Docked with *Salyut-6* on 19 September 1980 and returned to Earth with the same crew on 26 September 1980
Cosmos-1210	1980-76-A	USSR (PLE)	19 Sept.	195 km 268 km	88.8 min 82.3°		Photographic reconnaissance satellite. Recovered on 30 September 1980
Cosmos-1211	1980-77-A	USSR (PLE)	23 Sept.	215 km 261 km	89.1 min 82.4°		Photographic reconnaissance satellite. Recovered on 4 October 1980
Cosmos-1212	1980-78-A	USSR (PLE)	26 Sept.	216 km 275 km	89.1 min 82.3°		Earth resources satellite. Recovered on 9 October 1980
Progress-11 modified *Soyuz* spacecraft without the descent section; mass at launch: 7 tonnes	1980-79-A	USSR (BAI)	28 Sept.	193 km 270 km	88.8 min 51.6°		Cargo-spacecraft. Docked with the *Salyut-6/Soyuz-37* complex on 30 September 1980. Made to re-enter the Earth's atmosphere on 11 December 1980
Cosmos-1213	1980-80-A	USSR (PLE)	3 Oct.	207 km 343 km	89.6 min 72.8°		Photographic reconnaissance satellite. Recovered on 17 October 1980
Raduga-6 (Statsionar-3) 3-axis stabilized satellite; mass: 5 tonnes; solar cells	1980-81-A	USSR (BAI)	6 Oct.	36 000 km geostationary orbit	1444 min (24 h 04 min) 0.4°	5.7-6.2 GHz (reception) 3.4-3.9 GHz (emission)	Carries apparatus for transmitting television programmes and multichannel radiocommunications
Cosmos-1214	1980-82-A	USSR (PLE)	10 Oct.	181 km 368 km	89.7 min 67.2°		Photographic film recovery reconnaissance satellite. Recovered on 23 October 1980
Cosmos-1215	1980-83-A	USSR (PLE)	14 Oct.	499 km 553 km	95.1 min 74.0°		Electronic monitoring satellite

Name	Designation	Country (agency)	Date	Altitude	Period / Inclination	Frequency	Description
Cosmos-1216	1980-84-A	USSR (PLE)	16 Oct.	209 km 404 km	90.3 min 72.9°		Photographic film recovery reconnaissance satellite. Recovered on 30 October 1980
Cosmos-1217	1980-85-A	USSR (PLE)	24 Oct.	642 km 40 165 km	726 min 62.8°		Early warning satellite
Cosmos-1218	1980-86-A	USSR (PLE)	30 Oct.	178 km 374 km	89.7 min 64.9°		High-resolution photographic reconnaissance satellite. Recovered on 12 December 1980
FLTSATCOM-4 3-axis stabilized hexagonal satellite; width: 2.44 m; overall height: 6.70 m; mass at launch: 1876 kg; mass in orbit: 1005 kg	1980-87-A	United States USN (ETR)	31 Oct.	35 033 km 36 237 km in geostationary orbit at 172° E	1428.4 min 2.5°	240-400 MHz band (communications) 2252.5; 2262.2 MHz 2.4 W (telemetry)	Government communication satellite providing 23 UHF communication channels and one SHF up-link channel. Fourth in a series of five satellites
Cosmos-1219	1980-88-A	USSR (PLE)	31 Oct.	205 km 353 km	89.7 min 72.9°		Medium-resolution photographic reconnaissance satellite. Recovered on 13 November 1980
Cosmos-1220	1980-89-A	USSR (BAI)	4 Nov.	432 km 454 km	93.3 min 65.0°		Ocean surveillance satellite
Cosmos-1221	1980-90-A	USSR (PLE)	12 Nov.	207 km 424 km	90.5 min 72.5°		Medium-resolution photographic reconnaissance satellite
SBS-1 mass: 550 kg	1980-91-A	United States SBS (ETR)	15 Nov.	in geostationary orbit at 106° W		14-12 GHz band	United States domestic communication satellite. First of three all-digital business communications satellites. Transmits point-to-point voice, data, facsimile and telex messages. Ten transponders
48th Molnya-1 hermetically sealed cylinder with conical ends; mass: 1000 kg; 6 solar panels	1980-92-A	USSR (PLE)	16 Nov.	640 km 40 651 km	736 min 62.3°	800 MHz band 40 W (emission) 1000 MHz band (reception) 3400-4100 MHz (retransmission of television)	Television and multichannel radiocommunications

Name	Designation	Country (agency)	Date	Altitude	Period / Inclination	Frequency / Power	Description
Cosmos-1222	1980-93-A	USSR (PLE)	21 Nov.	624 km / 659 km	97.4 min / 81.2°		Electronic monitoring satellite
Soyuz-T3	1980-94-A	USSR (BAI)	27 Nov.	253 km / 271.5 km	89.6 min / 51.6°		For the first time in nine years three cosmonauts were launched aboard a *Soyuz*: L. Kizim, flight commander; O. Makarov, flight engineer; G. Strekalov, research engineer. *Soyuz-T3* docked with *Salyut-6* on 28 November and the crew boarded *Salyut-6* on 29 November. Soyuz-T3 was returned to Earth with its crew on 10 December, landing in Kazakhstan
Cosmos-1223	1980-95-A	USSR (PLE)	27 Nov.	614 km / 40 165 km	726 min / 62.8°		Early warning satellite
Cosmos-1224	1980-96-A	USSR (PLE)	1 Dec.	209 km / 403 km	90.3 min / 72.9°		Medium-resolution photographic reconnaissance satellite. Recovered on 15 December 1980
Cosmos-1225	1980-97-A	USSR (PLE)	5 Dec.	967 km / 1041 km	105.0 min / 82.9°		Navigation satellite
Intelsat-V F2 3-axis stabilized satellite: height: 6.60 m; mass at launch: 1950 kg; 2 solar arrays (1.2 kW)	1980-98-A	International INTELSAT (ETR)	6 Dec.	in geostationary orbit at 335.5° E		2202.5 MHz 3.5 W 5764 MHz 1 W (telemetry) 4-6 GHz (communications)	INTELSAT commercial telecommunication satellite: 12 000 telephone channels and two colour television channels
Cosmos-1226	1980-99-A	USSR (PLE)	10 Dec.	982 km / 1025 km	105.0 min / 83.0°		Navigation satellite
No name	1980-100-A	United States USAF (WTR)	13 Dec.	250 km / 39 127 km	63.8°		Satellite data systems spacecraft. Provides UHF communications and relays data and communications between satellite control facility earth stations
Cosmos-1227	1980-101-A	USSR (PLE)	16 Dec.	209 km / 325 km	89.5 min / 72.9°		Medium-resolution photographic reconnaissance satellite
Cosmos-1228 to Cosmos-1235 mass: 40 kg each	1980-102-A to 1980-102-H	USSR (PLE)	24 Dec.	1415 km / 1491 km	114.6 min / 74°		Government communication satellites

ITU Telecommunication Journal

Code name Spacecraft description	International number	Country Organization Site of launching	Date	Perigee Apogee	Period Inclination	Frequencies Transmitter power	Observations
Prognoz-8 pressurized central body: 4 solar panels	1980-103-A	USSR (BAI)	25 Dec.	550 km 190 000 km	95 h 23 min 65°		Automatic satellite to study influence of solar activity on Earth's magnetosphere. Experiments or equipment have been supplied by Czechoslovakia, Poland and Sweden
Ekran-6 **(Statsionar-T)**	1980-104-A	USSR (BAI)	26 Dec.	35 554 km 35 554 km geostationary orbit	1424 min 0.4°	5.7-6.2 GHz (reception) 3.4-3.9 GHz (emission)	Television relay satellite
Cosmos-1236	1980-105-A	USSR (PLE)	26 Dec.	180 km 388 km	89.8 min 67.1°		High-resolution photographic reconnaissance satellite

ITU Telecommunication Journal

Appendix D
LNA Noise
Temperature to Noise Figure
Conversion Chart (Approximations)

Noise Temp. in Degrees K.	Noise Figures in dB
69	0.825
70	0.95
75	1.0
80	1.05
85	1.125
90	1.175
100	1.3
110	1.4
120	1.5
130	1.6
140	1.7
150	1.8
160	1.9
170	2.0

Appendix E
Wind Pressure Table

Wind Speed in Miles Per Hour at +60° Fahrenheit	Pressure lbs./sq. ft.
20	1.6
21	1.75
22	1.925
23	2.1
24	2.3
25	2.5
26	2.7
27	2.9
28	3.2
29	3.4
30	3.6
31	3.9
32	4.1
33	4.4
34	4.65
35	4.95
36	5.2
37	5.5
38	5.8
39	6.0
40	6.5
41	6.75
42	7.0
43	7.5
44	7.75
45	8.0
46	8.5
47	8.75
48	9.25
49	9.75
50	10.0
60	14.5
70	20.0

Appendix F
Common Abbreviations

ac	Alternating Current
AM	Amplitude Modulation
AZ	Azimuth
C	Capacitance or Capacitor
CATV	Cable Television
C/I	Carrier-To-Noise Ratio
C/N	Carrier-To-Noise Ratio
dB	Decibel
dBw	Decibels Relative to One Watt
dc	Direct Current
EIRP	Effective Isotropic Radiated Power
EL	Elevation
FM	Frequency Modulation
G_A	Antenna Gain
GHz	Gigahertz (1000 Megahertz)
G_{LNA}	Low Noise Amplifier Gain
Hz	Hertz (1000 cycles per second)
kHz	Kilonertz (1000 Hertz)
K	Kelvin
Lat	Latitude
LNA	Low Noise Amplifier
Long	Longitude
Ma	Milliamperes (1/1000 ampere)
MATV	Multiple Access Television
MHz	Megahertz (1000 Kilohertz/1,000,000 Hertz)

Mos	Modulator
P	Power
P_{LNA}	Power Ratio Gain of Low Noise Amplifier
R	Resistance
rf	Radio Frequency
Rx	Receiver
SAT	Satellite
SAT_{lat}	Satellite Latitude
SAT_{long}	Satellite Longitude
S/N	Signal-To-Noise Ratio
T_A	Antenna Noise Temperature
T_{LNA}	Low Noise Amplifier Noise Temperature
T_{Rx}	Noise Temperature of Satellite Receiver
TVRO	Television Receive Only
Tx	Transmitter
V	Volts or Voltage
vswr	Visual Standing Wave Ratio

Appendix G
Noise Temperature
to Noise Figure Table

Conversion of noise temperature to dB noise figure or noise figure to temperature:

$$T = \left[\left(\text{antilog } \frac{dB}{10} \right) - 1 \right] 290 \qquad dB = 10 \log \left[\frac{T}{290} + 1 \right]$$

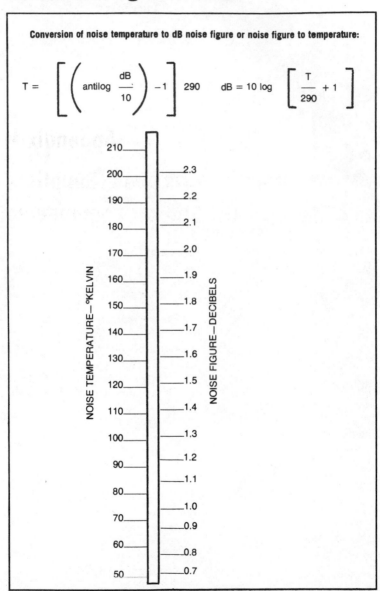

Appendix H
Manufacturers and Suppliers
of TVRO Earth Station Components

Avantek, Inc.
3175 Bowers Avenue
Santa Clara, CA 95051
(408) 727-0700

Channel One, Inc.
Willarch Road
Lincoln, MA 01773
(617) 259-0333

Channel Master
Division of Avnet, Inc.
Ellenville, NY 12428

Comtech Antenna Corp.
3100 Communications Blvd.
St. Cloud, FL 32769
(305) 892-6111

Downlink, Inc.
30 Park Street
Putnam, CT 06260
(203) 928-3654

Earthstar Corp.
Box 68
Steger, IL 60475
(312) 755-5400

Gardiner Communications Corp.
1980 S. Post Oak Road
Houston, TX 77056
(713) 961-7348

H and R Comm./Star View Systems
Route 3
P.O. Box 103G
Pocahontas, AR 72466
(800) 643-0102

Hamilton Satellite Systems, Inc.
1101 E. Chestnut, Ste. A
Santa Ana, CA 92701
(714) 543-5217

Heath Company
Benton Harbor, MI 49022
(800) 253-0570

Hero Communications
1783 W. 32nd Place
Hialeah, FL 33012
(305) 887-3203

Hi-Tek Satellite Systems
177 Webster Street
Route A455
Monterey, CA 93940
(408) 372-4771

ICM (International Crystal Mfg.
    Co., Inc.)
10 North Lee
Oklahoma City, OK 73102
(405) 236-3741

KLM Electronics, Inc.
Box 816
Morgan Hill, CA 95037
(408) 779-7363

Long's Electronics
P.O. Box 11347
Birmingham, AL 35202
(800) 633-3410

Robert A. Luly & Associates, Inc.
Box 2311
San Bernardino, CA 92406
(714) 888-7525

Microdyne Corp.
Box 7213
Ocala, FL 32672
(904) 687-4633

Microwave Associates
  Communications Co. (MA/COM)
South Avenue
Burlington, MA 01803
(617) 272-3000

National Microtech
P.O. Box 417
Grenada, MS 38901
(800) 647-6144

Satelco, Div. of Thunder-Compute
5540 W. Pico Blvd.
Los Angeles, CA 90019
(213) 931-6274

Satfinder Systems
6541 E. 40th Street
Tulsa, OK 74145

Satvision
Box 1490
Miami, OK 74354
(918) 542-1616

SED Systems, Ltd.
Box 1464
2414 Koyl Avenue
Saskatoon, Sask.
Canada S7K 3P7
(306) 664-1825

Starvision Satellite TV
Box 8624
Orlando, FL 32856
(305) 851-8332

Alan Swan
614 Cimarron
Stockton, CA 95210
(209) 948-5254

Third Wave Communications
3373 Oak Knoll Drive
Brighton, MI 48116
(313) 227-2822

Tri-Star General
4810 Van Epps Rd.
Brooklyn Heights, OH 44131
(216) 459-8535

Vitalink Communications Corp.
1330 Charleston Road.
Mountain View, CA 94043

Western Satellite
1660 Lincoln Street
Suite 2918
Denver, CO 80203
(800) 525-1636

Wilson Microwave Systems, Inc.
4286 S. Polaris Avenue
Las Vegas, Nevada 89103
(702) 739-7401

Winegard Company
Satellite Communications Division
3000 Kirkwood Street
Burlington, IA 52601

Appendix I
Satcom I Transponder
Frequencies and Polarization

Transponder	Frequency	Polarization
1	3720	V
2	3740	H
3	3760	V
4	3780	H
5	3800	V
6	3820	H
7	3840	V
8	3860	H
9	3880	V
10	3900	H
11	3920	V
12	3940	H
13	3960	V
14	3980	H
15	4000	V
16	4020	H
17	4040	V
18	4060	H
19	4080	V
20	4100	H
21	4120	V
22	4140	H
23	4160	V
24	4180	H

Antenna pointing angles, for Front Royal, Virginia

Earth station latitute (Deg., Min., Sec.)   =   38   55   0.00   North

Earth station longitude (Deg., Min., Sec.) =   78   10   0.00   West

Satellite	Longitude	Azimuth	Elevation	Hour Angle	Declination
EAST U.S. LIMIT	70.00	167.13	44.15	-9.25	-6.12
COMSTAR III	87.00	193.90	44.01	10.00	-6.12
WESTAR III	91.00	199.93	43.00	14.52	-6.11
COMSTAR II	95.00	205.72	41.66	19.03	-6.09
WESTAR I	99.00	211.21	40.01	23.52	-6.07
ANIK A1	104.00	217.62	37.58	29.11	-6.04
ANIK A2	109.00	223.54	34.80	34.66	-6.00
ANIK A3	114.00	228.98	31.73	40.18	-5.96
SATCOM II	119.00	233.99	28.43	45.65	-5.92
WESTAR II	123.50	238.16	25.31	50.53	-5.87
COMSTAR I	128.00	242.07	22.08	55.38	-5.83
SATCOM I	135.00	247.68	16.88	62.83	-5.75
WEST U.S. LIMIT	150.00	258.35	5.39	78.44	-5.58

(Courtesy of Comtech Antenna Corp.)

Appendix K
The Frequency Spectrum

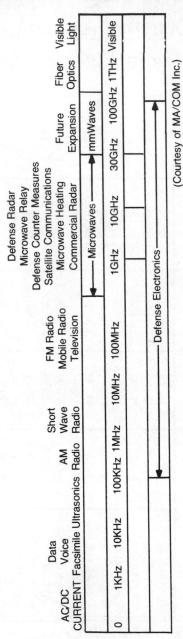

(Courtesy of MA/COM Inc.)

Appendix L
Preliminary EIRP Contour Map

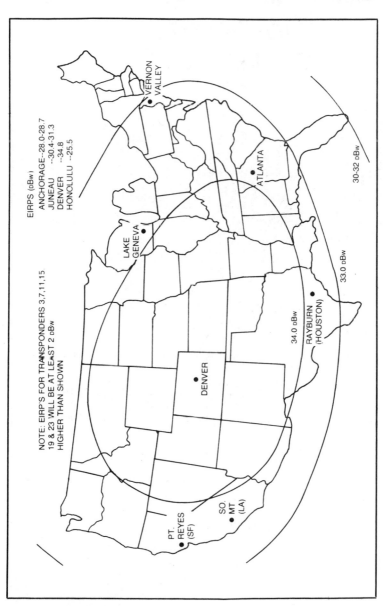

NOTE: EIRP'S FOR TRANSPONDERS 3,7,11,15
19 & 23 WILL BE AT LEAST 2 dBw
HIGHER THAN SHOWN

EIRPS (dBw)

ANCHORAGE--28 0-28.7
JUNEAU   --30.4-31.3
DENVER   --34.8
HONOLULU --25.5

VERNON
VALLEY

ATLANTA

LAKE
GENEVA

30-32 dBw

33.0 dBw

34.0 dBw

RAYBURN
(HOUSTON)

DENVER

PT.
REYES
(SF)

SO.
MT
(LA)

Appendix M
Preliminary Carrier/
Noise Vs EIRP Chart

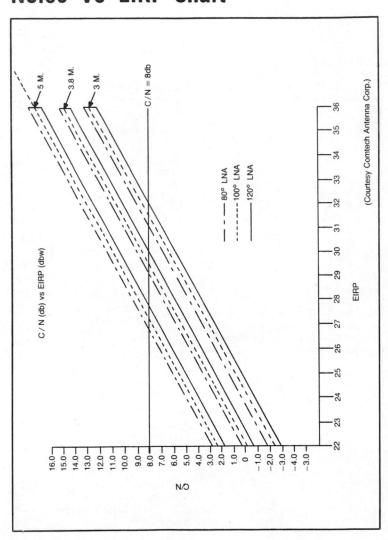

Index

A

Abbreviations	284
Active satellite system	18
ANIK B(2)	266
Antenna pointing angles	291
Antennas	78
active elements in	79
directive elements in	79
helix	81
helix array horn	64, 83
multi-element	77
parasitic elements in	79
quad	77
reflective elements in	79
spherical	68
yagi	80
ANIK 3	266
Avantek AR1000 receiver	147
Avantek AWC-4200 series	125
Avantek Line Extender	169
Avantek LNA/downconverter	130
Avcom Com 3 satellite receiver	168
Azimuth angles	205

C

Cable television system	210
Carrier noise vs EIRP chart	294
Checkout procedure	258
Coaxial cable	62, 215
Computer printout	196
COMSTAR 1	266
COMSTAR 2	264
Comtech model 650 receiver	152
Comtech 3-meter antenna	119
Comtech 3-meter antenna specifications	121
Comtech 5-meter antenna	120

Comtech 5-meter antenna specifications	123

D

Downlink D-2X receiver	140

E

Earth station assembly	254
Earth station block diagram	26
Earth station components, manufacturers	287
EIRP contour map	293
Equipment selection	255

F

Feedhorns	29
Frequency spectrum	292

G

Geostationary satellites, list of	267
Geosynchronous orbits	193
Gillaspie & Associates	137
Gillaspie Ampli-Splitter	166
Gillaspie FET bias circuit	225
Gillaspie LNA	221
Gillaspie LNA specifications	223
Gillaspie single conversion mixer	228

H

Hi-Tek conversion mixer	227
Horn antennas	64, 65
HR-100 satellite receiver	144

I

ICM Signal Purifier	169
ICM tunable audio	168
ICM TV-440 satellite receiver	152
Interference	261

K

KLM audio demodulator	237
KLM Electronics	
Sky Eye I satellite receiver	167
KLM first i-f amp	235
KLM mixer module	234
KLM receiver alignment	241
KLM receiver block diagram	234
KLM second i-f amp	236
KLM video demodulator	236

L

Line of sight broadcasts	12
LNA noise temperature to	
noise figure conversion chart	282
LNA printed circuit board	222
Low noise amplifier	124

M

Magnetic variation	206
Manufacturers and suppliers	287
Microdyne Corporation	142
Microwave Associates	105
antennas	100
3-meter antenna	105
3.7-meter antenna	108
4.6-meter antenna	110
specifications	107, 109, 111
VR-3X satellite receiver	153
VR-4XS satellite receiver	158
Microwave Associates LNA	130
Modulators	124
Multi-hop transmission	16
Multiple access ground stations	184

N

Noise temperature to	
noise figure table	286

P

Passive reflector	14
satellite system	17
Parabolic dishes	55
/spherical comparison	74
Paraboloid reflectors	65
Programs, satellite TV	263
Protractor, use of	207

Q

Questions and answers	39

R

Radio waves	12
Receivers	124
Receiving stations	10
Remote control	37, 87

S

SATCOM 1	263
2	266
1 transponder frequencies	290
SatVision model SV-11	118
Satellite Supplies Inc.	144
Satellites launched in 1980	270
locations	291
location techniques	193
receiver	29
reception printout	196
TV programs	263
windows	43
Sighting device	245
Site selection	244
Skyview I antenna	70, 103, 113
Skywave communications	13
Sky Eye II receiver kit	231
specifications	232
Soldering	239
Space communications	15
Spherical antennas	68
/parabolic comparison	74
Surplus components	54

T

TVRO antenna system	27
modulator	31
receiver kit	230
systems	175
Teleco Inc. Earth station dish	211
low noise amplifier	213
Templates	100
Third Wave	
Communications Corp.	189
Transmitting stations	8

W

Waveguides	61
rectangular	63
WESTAR 1	265
2	265
3	265
Wind breaks	251
pressure table	283
Winegard Earth station antennas	116
Wilson 3.35 meter antenn	86
4 meter antenna	88
Wilson antenna parts list	90
Wilson antenna support frame	95
Wilson Microwave Systems	85
Wilson rotor mounting plate	97

V

VR-3X satellite receiver	153
Vitalink Communications	139

Edited by Roland Phelps